The Inner Journey

How to navigate Life's Challenges with Confidence and Grace

NEW LIFE
CLARITY
PUBLISHING

New Life Clarity Publishing
205 West 300 South, Brigham City, Utah 84302
Http://newlifeclarity.com/

Printed in the United States of America
ISBN- 978057896999
Copyright@2024 Loredana Climena Stupinean

A Love Letter dedicated to my family

—◆◆◆◆◆◆—

To my dearest children, Isis Veronica and Dominick. This book is part of my legacy to you—a gift woven from the wisdom and teachings I have gathered along my journey. May it serve as a guiding light, a source of inspiration, and a reminder of my infinite love for you.

As I write these words, know that you are the very essence of my being, the light of my life, and my greatest source of inspiration. Every moment with you is a treasure beyond measure, and I am endlessly grateful for the joy and meaning you bring into my life. You are both extraordinary—filled with potential, creativity, and an infinite ability to shape the world in your own unique way.

I have no doubt that you will accomplish remarkable things. Always remember that life is a magnificent journey, filled with both triumphs and challenges. When obstacles arise, stand tall and meet them with courage, knowing that the strength to overcome anything already exists within you. Follow your dreams with passion, trust the voice of your heart, and always choose what makes your soul vibrate.

You hold the power to shape your own destiny, and I believe in you—today and always.

To my beloved husband, your love and unwavering support have been my anchor in every storm. Thank you, my love, for standing by my side, for believing in me even when I struggled to believe in myself, and for walking this journey of life hand in hand with me. Your strength lifts me, your encouragement fuels me, and I am profoundly grateful to have you as my partner in this grand adventure.

To my beloved mother, your love has been a guiding light throughout my life, illuminating my path with wisdom, kindness, and an unshakable belief in me. Your presence has been my shelter, your words my greatest encouragement, and your heart the place where I have always found comfort.

Thank you for teaching me the power of resilience, for lifting me when I stumbled, and for always guiding me toward the fulfillment of my dreams. Your love has been the foundation of our family, the warmth that has kept us together, and the gift I carry with me every single day. No matter where life takes me, I will forever hold your lessons, your love, and your light deep within my heart.

To my beloved father, though you are no longer physically here, your presence lives on in our hearts, in the strength you instilled in me, and in the unwavering values you passed down.

Thank you for teaching me what it means to be strong—to remain steady in the face of the winds and storms of life, to walk with integrity, and to never waver from what I believe in. You led not through words, but through the power of your example, showing me that true resilience comes from within. I will always carry your wisdom in my heart, and I assure you of my forever love and respect.

Together, you are the greatest blessings of my life. I hope this book stands as a testament to my love for you all—a legacy of encouragement, wisdom, and dreams fulfilled. May it remind you always that you are never alone; love surrounds you, guiding you through every chapter of your journey. As you read these words, feel my heart with you—celebrating your victories, comforting you in your struggles, and cheering you on as you step boldly into your destiny.

Embrace love, cherish every moment, and never forget that no matter where life takes you, you are deeply, profoundly loved.

Foreword

———·◆◆◆◆·———

The Inner Journey. How to Navigate Life's Challenges with Confidence and Grace is more than just a personal development book; it is a scientifically grounded compendium that brings together the latest discoveries about life, the human being, and the complexities of the human body. Loredana Climena Stupinean offers a well-researched and masterfully crafted work that blends modern concepts from fields such as epigenetics, neuroscience, and neuro-linguistic programming (NLP) with profound reflections on human nature and spirituality.

This volume is a comprehensive and well-argued exploration of human potential, providing readers with cutting-edge insights into how thoughts, emotions, and the environment influence genetic expression, health, and overall well-being. Loredana presents complex ideas in a clear and accessible manner, such as how the brain functions as an energy receiver-transmitter and how our perceptions and paradigms shape our inner and outer realities. The book excels in balancing rigorous scientific documentation with practical application.

Each chapter is structured to be not only informative but also transformative. Readers are guided through practical exercises in introspection, visualization, and mindfulness techniques, offering concrete tools to integrate the knowledge into their daily lives. This unique combination of theory and practice makes this book an essential guide for anyone seeking to discover and maximize their potential.

Another notable strength of the book is the way the author addresses broad themes, from the spiritual essence of human beings to the complex functions of the physical body, exploring topics such as interconnectedness with the environment, the law of attraction, and the impact of habits on mental and physical health. These themes are supported by scientific studies and concrete examples, which lend credibility and depth to the text.

Loredana Climena Stupinean is not just an author, she is a riguros professional who skillfully translates complex information into an accessible language without compromising the scientific rigor. Her writing is clear, elegant, and profound, creating a genuine connection with the reader. Every line reflects her dedication to her mission of guiding others in discovering their authentic selves and achieving inner balance.

The Inner Journey. How to Navigate Life's Challenges with Confidence and Grace stands out for its density of information and the clarity with which it explains the subtle mechanisms governing life and human existence. It is a work that enriches not only the mind but also the soul, offering a comprehensive framework for personal and professional transformation.

I highly recommend this book to anyone seeking a deeper understanding of the human being as well as practical tools for successfully navigating life's challenges. This volume offers an indispensable guide for anyone who wishes to enhance self-awareness and build a more conscious and harmonious life. It is a reference volume, written with authentic passion and a profound respect for both science and spirit, deserving to be explored and revisited time and again. The Inner Journey. How to Navigate Life's Challenges with Confidence and Grace is a book that should not be missing from the personal library of anyone striving to be the best version of themselves every single day!

By Professor BAKHAYA IRINA MIHAELA
"Alexandru Ioan Cuza" Police Academy Bucharest, Romania

Introduction

————•◆◆◆◆◆•————

Dear reader, first of all, thank you for choosing this book. Your decision to pick it up means a lot to me, and I appreciate the time and attention you're giving to these pages. Writing this book has been a labor of love, a journey of discovery and reflection, and I'm excited to share it with you. Each chapter is filled with thoughts, experiences, and insights that I hope resonate with you on a personal level. As you immerse yourself in these words, know that you are not just a reader; you are a part of this journey, and I am grateful for your presence. I hope this book sparks inspiration, provokes thought, feelings, and insights and perhaps even offers a sense of connection as we explore these ideas together. Thank you for being here.

In this world teeming with uncertainties and challenges, the quest for inner strength and resilience has never been more vital.

"Inner Journey. How to Navigate Life Challenges with Confidence and Grace" serves as a guiding light for those seeking to transform adversity into opportunity, Ego into Love and Compassion, challenges into experiences. This book invites readers to embark on a profound exploration of their inner selves, uncovering the tools and strategies necessary to face life's obstacles with poise and assurance.

Through a blend of scientific information, discoveries and proofs, practical exercises, and timeless wisdom, you will discover how to cultivate a mindset that embraces change, fosters self-compassion, and nurtures a sense of purpose and how to apply it in your life for wellbeing. Each chapter is designed to empower you to confront fears, overcome setbacks, and emerge from life's trials with renewed strength and clarity. As you navigate the winding paths of your journey, you'll learn that true confidence and grace stem not from the absence of challenges, but from the courage to meet them head-on.

Join me on this transformative journey and unlock the potential within you and face life's challenges with unwavering confidence and grace.

Table of Contents

CHAPTER 1

————— ·◆◆·◆·◆◆· —————

Discovering your Self-Awareness: A Journey Within.

What is the Power of Self-Awareness and how to use it

"NOSCE TE IPSUM - GNOTHI SEAUTON - KNOW THYSELF"

WHY WOULD THE ANCIENT PHILOSOPHERS INSIST SO MUCH UPON THIS? ....

Ancient philosophers, such as Socrates, Plato, and Aristotle, emphasized the importance of knowing oneself because they believed that self-knowledge was the key to living a fulfilling and virtuous life. When we understand our own strengths, weaknesses, desires, and motivations, we can make better decisions, cultivate our virtues, and live in accordance with our true nature. Self-knowledge was also seen as a way to achieve inner harmony and peace of mind. By being aware of our own thoughts and emotions, we can manage them and avoid being controlled by them.

Self-awareness could also lead to greater empathy and understanding of others, fostering better relationships and social harmony. Furthermore, understanding oneself was seen as a path to personal growth and self-improvement. Becoming aware of our flaws and limitations gives us the possibility to work on overcoming them and striving to become better versions of ourselves. This process of self-reflection and self-improvement was seen as essential for achieving excellence and contentedness in life.

The ancient philosophers insisted on knowing oneself because they believed it was the foundation for living a virtuous, harmonious, and fulfilling life. Self-knowledge was seen as the key to understanding one's true nature, achieving inner peace, fostering better relationships, and striving for personal growth and excellence.

Self-knowledge was considered by ancient philosophers to be the foundation of all other forms of knowledge. They believed that understanding oneself was essential for gaining insight into the world around us and making sense of our place in it. By knowing our own biases, beliefs, and limitations, we can better navigate the complexities of life and make more informed decisions.

Additionally, self-knowledge was seen as a way to cultivate virtues such as wisdom, courage, temperance, and justice. By understanding our own values and beliefs, we can align our actions with our moral principles and lead a virtuous life. This alignment between our thoughts, emotions, and actions was believed to bring about a sense of inner harmony and fulfillment.

Furthermore, self-knowledge was seen as a form of self-empowerment. Coming to understand our own strengths and weaknesses allows us to play to our strengths and work on improving our weaknesses. This process of self-improvement was seen as essential for personal growth and development, as well as for achieving our full potential.

Becoming aware of who we are is a transformative and empowering process that involves deep introspection, self-reflection, and a willingness to explore our inner landscape. It requires a commitment to self-discovery, a willingness to confront our fears and insecurities, and a desire to understand our true selves on a deeper level.

I liked very much what one of my masters said: he compared our existence/ life with an American university where the student has a set of mandatory exams and some optional! Well, according to his saying, we are all here to pass the mandatory ones and those of us who go for the optional ones they upgrade to the next level!

In nowadays world filled with so many distractions and external influences, it is so easy to lose sight of who we truly are and what we stand for. The inner journey to self-discovery is a deeply personal and transformative process that requires courage, introspection, and a willingness to explore our inner selves. It is a journey that can lead to greater self-awareness, authenticity, and fulfillment in our lives.

This book is a guide to embarking on your own journey to self-discovery, uncovering the layers of your identity, and gaining insights into your true self. Through reflections, exercises, and practical tips, you will gain insight on how to cultivate self-awareness, embrace vulnerability, and align your actions with your values and beliefs. Becoming more aware of who you are, gives you the key to unlock your full potential, overcome obstacles, and live a life that is true to yourself.

Join me on this transformative journey of self-discovery and empowerment. Let's explore the depths of our inner selves, uncover our true essence, and awaken the power of self-awareness within us. The path to self-knowing begins here – are you ready to embark on this life-changing adventure?

So, let us dive into the depths of this inner journey of self-discovery and how to become aware of our true nature using some practical tools.

Questions like "What are we? "Who are we?" have made many years the subject of philosophical but also existential debates and are still doing so even nowadays! We are extraordinary beings having physical experience, having the most advanced and complex "machine" as a body! Our body is equipped with five senses (visual, hearing, kinesthetic, olfaction and taste) and through these we experience life and consequently what we define as our reality. If we are to do a small imaginative exercise, we can envision us wearing a pair of glasses and on top of them adds one sense after another, so that these act as filters for our vision! To all these, of course, it adds our own paradigm: all the ancestral programs, the society programs, the school, the friends' programs etc. so that our perception becomes actually the result of all these filters! But are these all that is? When we look in a mirror and see our reflection -a shape - our body, we tend to identify ourselves with that image as being us and we become attached to it! But is it really what it is? Are we just a body or a name? Many scholars, philosophers and research have come to the conclusion that we are definitely more that this stating that: **we have a body**, **we have a name** but **we are neither one nor the other**! **We are divine essence, Spirit, having a physical experience**!

"We are born at a given moment, in a given place and, like vintage years of wine, we have the qualities of the year and of the season in which we are born" said Carl Gustav Jung, the father of modern psychology. We are the result of our environment and how we interact with it! In other words, what epigenetics is also stating: the environment is dictating our behavior in its totality, our values and principles as well as our health and wellbeing.

Since the day we are born we are continuously taught our names, the mother tongue -the language we speak, the national identity and we get attached to all that, believing that we actually are all that! Our parents, grandparents, friends, all the educational system constantly programs us to become a member of the society we are born into. So that by the time we are grownups we form our paradigm as a reflection of this "collection" but without being necessarily aware of its influence in shaping our lives and we end up calling as our reality. Maybe some of you have heard the question "what do you want do become when you grow up?" during childhood, as if we are not already an entity – the most developed of all living beings on Earth – a Human. Well, this question is an example of how far from understanding our true nature society was and continues mostly to be by judging us and dictating our direction in life. Bearing in mind our starting point, that *we are a pure essence – Spirit – having human experience*, it becomes obvious that the previous question makes no sense at all! Whereas questions like, *what would you like to do* or *what would you like to explore* are more helpful and valuable. Some of us know intuitively even from birth their natural talents, their call and they follow like a red thread, step by step their life purpose or mission.

Others, you may say, not that lucky, do not realize exactly what are their talents or even if they have any and are totally confused, spending years to realize what are they born to do, what direction to go to. And yet, they manage through ups and downs to make something out of their life, sooner or later! But some are looking deeper and deeper and further looking to understand what it with them in this world is, on this planet and have a difficult time in finding their purpose.

Obviously, there are debates regarding the idea of a 'life mission' but even if there are believes denying it, the mere *birth* and *death* concepts become redundant if there would be no purpose or mission in this life experience, don't you think? Even cases like miscarriages, babies that die at birth or shortly after, as challenging as these events are, each has in itself a purpose!!! ***Nothing happens without a reason!***

The author Dolores Canon, a self-described "past-life regressionist" and hypnotherapist who specialized in the "recovery and cataloging" of "Lost Knowledge" in her volumes "Convoluted Universe" (vol.I-III)*[1] describes **three waves of volunteers,** who are coming to assist and help the world to heal and go through its transition. One of these waves includes the ones that came here for the mere purpose of being in physical form, working mostly behind the scenes! So, not everyone is here TO DO something, some are here just to be in these human bodies, like beacons of light, inspiring others by their simple presence and that's their mission! All is not necessary, it seems!

Now, this is a whole science that developed, some claim, far before our existence, and it studies the human holistically, following a mental map based on the birth coordinates of the individual. A very complex and comprehensive system was born or more accurate was gifted to mankind, more precise than astronomy, astrology or numerology! A quintessence of all systems under one roof: it was named Almuza!

Since ancient times humans have been preoccupied to understand our very own nature of being humans and they understood that who we are is far beyond the limitations of a physical form and all the sensations perceived through the senses. All our thoughts, emotions, feelings, values, beliefs and unique experiences are aspects that shape our identity and contribute to our overall understanding of who we are!

But let's continue our journey a little bit on the road of scientific discoveries.

As we all know and science has already stated, humans are the most intelligent beings on this planet, so far and there is no doubt about it! It is believed that we have appeared or better said have

---

[1] The volumes *Convoluted Universe* are presenting and explaining the planetary transformations and the transition of Earth as a planet together with humanity as earthlings, from a three-dimensional reality to a forth and a fifth dimensional reality. For those curious or interested to find out more about this subject I really recommend the lecture of these volumes.

been created by *a perfect design*, thousands of years ago, meaning, *we are humans by design* - as Gregg Braden very well stated in his book with the same title, perfect divine creatures living in a body, having a name, having a life-path, same way as we have a car, a job and so on.

"The modern human, or Homo Sapiens, appeared to have a perfect design" (Bronfenbrenner, 1979). And the most distinguishing quality that sets us, humans, *Homo Sapiens,* apart from other species are our abilities **to think abstractly** and **to engage in complex cognitive processes**. We are conscious beings capable of introspection, self-questioning our own nature and purpose in life! Every single one of our thoughts triggers a specific emotion and together these give birth to feelings; our mindsets help us shape our perceptions, our attitudes and our behaviors.

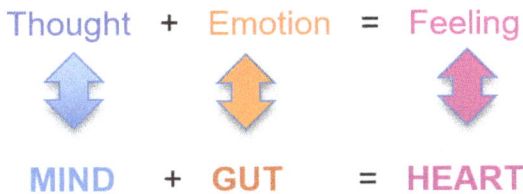

Thought + Emotion = Feeling

MIND + GUT = HEART

Latest scientific discoveries confirm that *there is no place in our brain to locate the mind* and that **actually our brains behave like a receptor-emitter of energy/thoughts**. The science of epigenetics reveals the superpower of your mind: our thoughts can change our genes, hence what Buddha said (***With our thoughts we create the world***. He got it much more accurately than Descartes *I think, therefore I am.* But more to follow in the next chapters.

But let's see what epigenetics stands for exactly: epi (above) & genetics (genes) is the study of gene expression, or the changes in how our DNA folds, thereby turning particular genes on or off. Like we define dermis and epidermis!

Dr. Brian Lipton[2] said that: *"genes do not turn on and off; genes have no self-control; genes are just blueprints. We now know that some action has to select a blueprint and read a blueprint – and all of a sudden, we realize that the genes do not control the action. The action is actually controlled by the environment, as it is also our perception of the environment. That action becomes very significant because if our genes and our life are controlled by our relationship to the environment, then we are not necessarily victims anymore. We are the ones, who can change the environment or change our perception of the environment, and in that case, we have control above the genes – we have epigenetics."*

Epigenetics revolutionizes the way we used to look at things and turns up to show how our behaviors and the environment can cause changes that affect the way our genes work. Unlike genetic changes, epigenetic changes are reversible and do not change our DNA sequence, but they can change how our body reads a DNA sequence.

Epigenetics could be pictured as a conductor and the genes as its orchestra! This tremendously important discovery actually freed us instantly from the belief that we are doomed if one of our parents or grandparents had a genetic disease.

After this discovery we came to realize that the *body* is created to function and adapt to the environment and the *mind* is the power to heal whatever disease without any human intervention! That being said, doctors, pharmacies, healers of all kinds etc. would simply not be necessary anymore, nobody would actually need them but only to advise or supervise at most!!! Do you understand

---

[2] Brian Lipton, PhD, is an internationally recognized biologist, author, and pioneer in the new science of epigenetics. Dr. Bruce Lipton's epigenetics research on cells since the 1970s has shifted the way he thinks, not only about biology but about life. His research has taught him that cells have intelligence and live in cooperative communities, working together in response to the signals they get from their environment. His research reveals more about genes and cell behavior than we ever knew before.

now the power of the mind and what we can do when we manage to master it?

The body cannot outrun the thoughts that flow through our mind, it is impossible, as example look at the kids when they play and talk about green dogs or pink horses or being princes or super-heroes, they just follow their own thoughts; every single one of our cells literally vibrates with the energy that our thoughts create. We create a perception of the reality we relate to and we call it **our reality**, based on our paradigm (which is the sum of all our thoughts, senses and all the programs we inherited as well as those we created all through our life). If the perception in our mind is reflected in the chemistry of our body, and if our nervous system reads and interprets the environment and then controls the blood's chemistry, this means that we can literally change the fate of our cells by altering our thoughts.

So, basically, the energy of our thoughts manifests our experiences. As such:

1. Positive thoughts manifest positive emotions and positive experience:

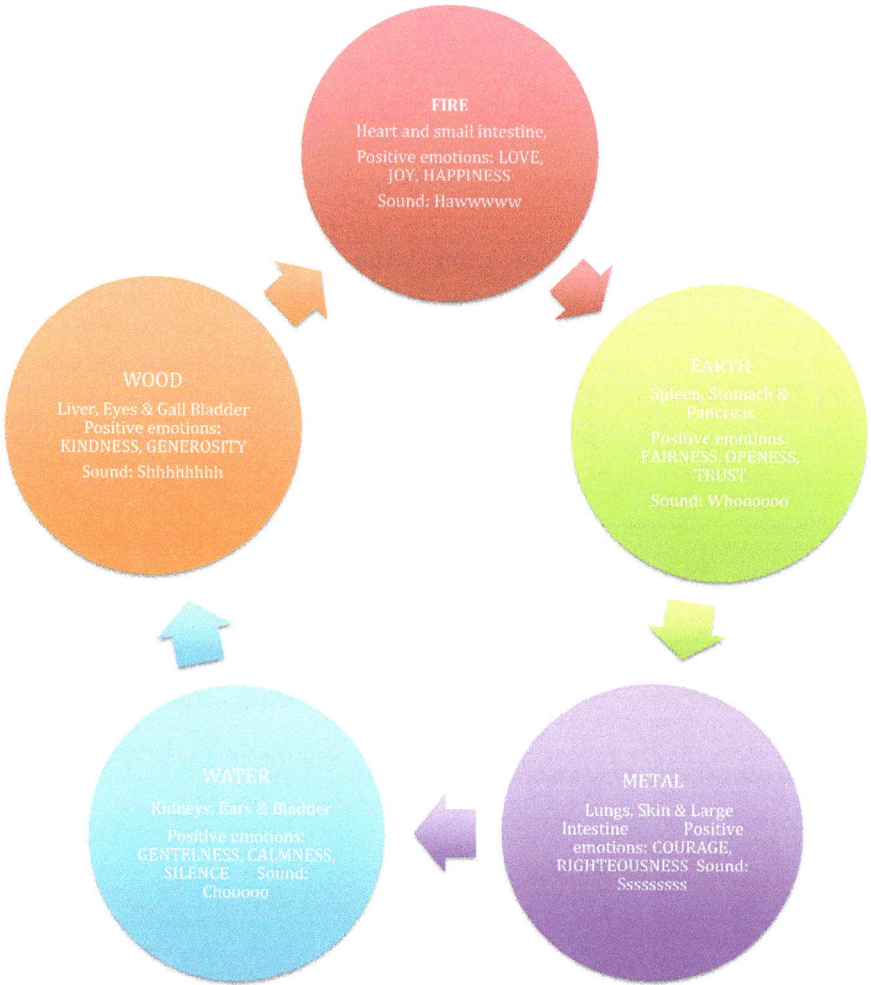

FIRE
Heart and small intestine,
Positive emotions: LOVE,
JOY, HAPPINESS
Sound: Hawwwww

EARTH
Spleen, Stomach &
Pancreas
Positive emotions:
FAIRNESS, OPENESS,
TRUST
Sound: Whooooooo

WOOD
Liver, Eyes & Gall Bladder
Positive emotions:
KINDNESS, GENEROSITY
Sound: Shhhhhhhh

METAL
Lungs, Skin & Large
Intestine   Positive
emotions: COURAGE,
RIGHTEOUSNESS Sound:
Sssssssss

WATER
Kidneys, Ears & Bladder
Positive emotions:
GENTELNESS, CALMNESS,
SILENCE   Sound:
Chooooo

2. Negative thoughts manifest negative emotions and negative experiences.

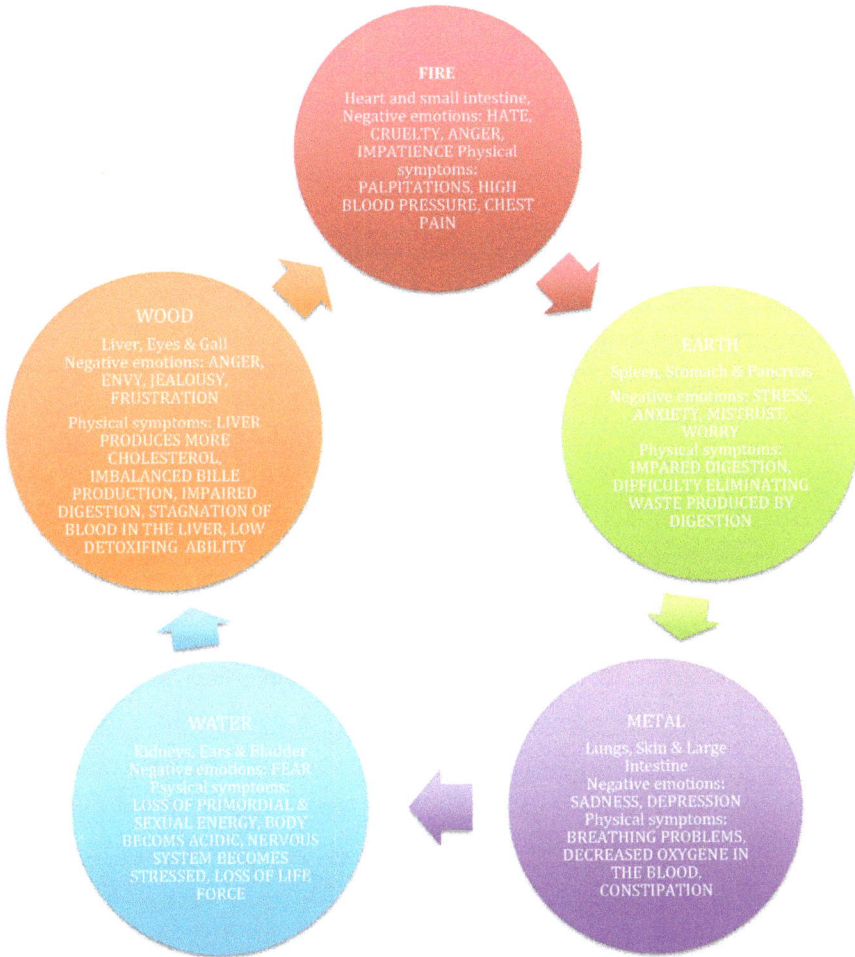

FIRE
Heart and small intestine,
Negative emotions: HATE,
CRUELTY, ANGER,
IMPATIENCE Physical
symptoms:
PALPITATIONS, HIGH
BLOOD PRESSURE, CHEST
PAIN

WOOD
Liver, Eyes & Gall
Negative emotions: ANGER,
ENVY, JEALOUSY,
FRUSTRATION

Physical symptoms: LIVER
PRODUCES MORE
CHOLESTEROL,
IMBALANCED BILLE
PRODUCTION, IMPAIRED
DIGESTION, STAGNATION OF
BLOOD IN THE LIVER, LOW
DETOXIFING ABILITY

EARTH
Spleen, Stomach & Pancreas

Negative emotions: STRESS,
ANXIETY, MISTRUST,
WORRY
Physical symptoms:
IMPARED DIGESTION,
DIFFICULTY ELIMINATING
WASTE PRODUCED BY
DIGESTION

WATER
Kidneys, Ears & Bladder
Negative emotions: FEAR
Physical symptoms:
LOSS OF PRIMORDIAL &
SEXUAL ENERGY, BODY
BECOMS ACIDIC, NERVOUS
SYSTEM BECOMES
STRESSED, LOSS OF LIFE
FORCE

METAL
Lungs, Skin & Large
Intestine
Negative emotions:
SADNESS, DEPRESSION
Physical symptoms:
BREATHING PROBLEMS,
DECREASED OXYGENE IN
THE BLOOD,
CONSTIPATION

"Like attracts like" is the central universal principle that makes up the law of attraction. Based on this law, similar things are attracted to one another.

We understand now that our thoughts have tremendous power as well as the ability to shape our life for either good or bad. For example, maybe in our childhood, we accept a thought someone told us, like for example: "You're worthless, you don't matter." If we accepted that thought, even though it was wrong, that very thought like "a magic wand", by repetition, over and over in our mind, it ended up shaping our life and creating future problems. We must understand that our thoughts create our own illnesses and as such we are the only ones able to heal them by starting to ask ourselves what is the root cause of our sufferance, and looking to find out what thoughts and emotions we were feeling? Therefore, it is important to learn to manage our thoughts and to become aware of them since they control our existence.

But what is that 'something' that's doing that? How can we manage to control our thoughts? Well, it is our consciousness what enables us to be aware: to perceive, interpret and find meanings of our environments and of our world. Becoming self-aware enables us to reflect upon our existence and ponder the deeper questions of life!

In summary, we are more than our physical bodies and their sensory experiences. This multifaceted nature makes us complex, ever-evolving individuals, with a unique sense of identity. Being unique sets us apart from one another and enriches the diversity of the world. Our unique attributes allow us to offer distinct contributions to society, whether through our creative endeavors, problem solving skills or compassionate actions. Embracing our uniqueness can lead to personal fulfillment and a sense of purpose. However, while distinct individuals, we are also part of something greater than ourselves: like the fingers of a hand that are part of a palm, which in turn is part of an arm, which is part of a body and so on! As individuals we are all interconnected, not only with other humans but also with all living beings and with the whole world around us. **Diversity enriches unity**! Being aware of our interconnectivity generates empathy, compassion, and a sense of responsibility towards others as well as towards the environment. By acknowledging that our thoughts, emotions, actions impact not only ourselves but also those around us, and even future generations, we actually become fully aware of ourselves! This **self-awareness** encourages us to act in ways that promote harmony, sustainability and justice. As such, the constant quest towards improvement and personal growth serves as the main ingredient for transformation. But only embracing both our uniqueness and our connection to the collective, to the whole, will open us the door to a balanced understanding of who we really are and of our place in the world!

Self-awareness is the key to conscious knowledge and understanding of our unique character, emotions, motives, and desires. It is the ability to introspect and reflect on our thoughts, feelings, and actions, allowing us to gain insights into our strengths, weaknesses, values, and beliefs. When we gain a proper understanding of who we are, we automatically acquire the ability to make decisions and take actions that are aligned with our true selves. If we want to be successful and whole then we must become self-aware of our aspirations, have clarity of our values, specify our purpose, and establish worthwhile objectives.

Self-awareness is the foundation of personal empowerment!

We can make decisions that are in line with our true selves and move closer to fulfillment when we are self-aware and clear about whom we want to be and what we want to achieve.

Self-awareness is equally essential to developing emotional intelligence. Understanding and managing our emotions will help us grow in empathy and improve our interpersonal relationships. This further enables us to interact with others more effectively, settle disputes, and forge healthier relationships.

Moreover, self-awareness empowers us to embrace our strengths and recognize areas for growth. By understanding our limitations, we can seek opportunities for personal development and self-improvement. It encourages us to step out of our comfort zones, take calculated risks, and embrace new experiences that contribute to our growth and evolution.

As human beings, self-awareness is especially crucial in navigating societal expectations, biases, and stereotypes. It enables us to challenge limiting beliefs, break free from societal norms, and redefine our definitions of success and happiness. By being self-aware, we can confidently assert our boundaries, make choices that serve our best interests, and forge our unique paths in life.

To harness the power of self-awareness, it is essential to cultivate mindfulness and engage in regular self-reflection practices. Keeping a journal, daily meditation, and seeking guidance from experts from the personal development area like life coaches, numerologists, Almuza consultants and so on can all contribute to deepening our self-awareness. Additionally, we can surround ourselves with a supportive community of like-minded individuals that can provide a proper nest for growth and self-discovery.

In conclusion, self-awareness is the main key to unlocking the full potential of each person. By understanding ourselves on a deeper level, we gain the power to navigate life's challenges with confidence and grace. Practicing, on a constant basis, self-improvement techniques and life coaching courses or therapies, we can embark on a transformative journey towards personal empowerment, embracing our authentic selves and becoming the true creators of our reality, living life on our terms.

CHAPTER 2

---◆◆ ◆ ◆◆---

Uncovering Your True Identity

Neuro-linguistic-programing (NLP) strategies

In our life journey it is very easy to lose sight of who we truly are while navigating through the demands and expectations of everyday life. We frequently find ourselves following social expectations and roles while ignoring our real passions and desires. We let ourselves be entrapped into the meanders of the world's society with all its distractions: social media, movies - Netflix etc. Hypnotized by all these distractions disguised as creative activities, we become the puppets in our own life puppet show consequently creating a puppeteer reality! But for those who want to be free, strong and whole, it's imperative to set themselves out on a quest to discover who they really are.

This section aims to provide some suggestions to empower you to embrace your authentic Self and live a life of your own, with confidence and grace by guiding you through the process of Self-discovery and Self-awareness.

You will be able to understand that there are ways and tools to unlock your true potential and the power to overcome life's obstacles by looking deep inside of yourself.

And here are some steps I propose to follow, dearest reader:

Step 1. Uncover your true identity.

In the pursuit of finding our true identity the first and most obvious question to ask is the famous one already:

Who I really am?

For a proper answer to be attained we must first reflect upon our values, passions, and aspirations, take the time to explore what truly matters to us and what brings us joy. This can be seen as a journey where we look to reconnect with our childhood dreams, to rediscover forgotten hobbies, or exploring new interests. By aligning our life with our core values and passions, we will find a sense of fulfillment and purpose that will propel us forward on our path.

A very clear example of how we can use all the parameters within the context of the neuro-linguistic-programing (NLP) for achievement of self-insight and awareness is to learn different strategies in order to gain the proper tools and insight to make better decisions and reach our goals.

SOURCE

VALUES/TRUTHS

HABITS AND BELIEVES

ACTIONS/BEHAVIORS

ENVIRONMENT

The main idea in NLP is that the *Source, Values, Habits & Beliefs, Behavior & Actions* and *Environment* are interconnected and influence each other.

So, here's a short analysis of each and how they relate:

Source

Our source or sources refer to the feedback we receive from the world around us, in the form of information, experiences, and knowledge. But we must pay very close attention and be careful of the quality and diversity of the sources we expose ourselves to, as they shape our perception of reality and further impact our decision-making. So, the Source must be of a high standard and well verified. Best follow our inner voice/intuition when choosing the adequate Source. It is not an easy task given the volume of information we have to deal with but, not an impossible one!

Suggestion:

If you want to connect with your inner voice or intuition, you can immerse yourself into a relaxing state of calm and tranquility by focusing for 2-5 minutes on your breath, while closing your eyes. Listening to soft alpha waves beats definitely helps. Many social media channels nowadays promote a really good quality of relaxing tones but rest aware when choosing the appropriate ones, better ask a specialist. What I can recommend from my own experience is the Heart Math Institute tunes, the Silva method alpha waves and the Equi Sync tunes.

*So, **when you want to calm and relax**, you can start by focusing on your breath and repeat in your mind, **while inspiring: "with each breath, I allow myself to relax and go deeper and deeper till I reach a profound state of calmness where I hear or feel clearly my intuition"** and on the **exhale release all your worries** saying: "I release all my doubts, fears, worries and discharge!"*

Then place your attention at the level of your heart chakra (on your chest), you can place your left hand on the specific place to better focus your attention cause "where attention goes, there energy flows!" The language of the heart is silence, the opposite of what we are used to when it is about communication. So, keep quiet, listen, trust yourself and follow the very first impulse it comes to you, without doubts! Be persistent and continue until you succeed. Repetition is the mother of learning and learning is the completion of a program setting, which can lead to building a new and healthier paradigm.

Values

Regarding our values, they are the underlying principles and beliefs that guide our actions and choices. So, you should allocate some time to reflect on what truly matters to you in

life: is it family, love, wellbeing, truth, happiness, freedom, or what is it. Try to identify your core values and ensure that your decisions and goals align with them. This will provide clarity and a sense of purpose and fulfillment. The secret here is to be able to identify the values because whether we know them consciously or subconsciously, they lead our lives so, we should better get to know them if we want to understand our life.

Suggestion:

    a. *When you want to analyze if compatible with a possible partner, in business or and relationship you can both write on a different piece of paper 3-5 core values that define each one of you and then compare. Define what these mean to each of you and then check your answers.*

    b. *If you want to achieve a goal: be very specific about your goal and identify what values determine and help you to achieve it; write it down at present tense as if already fulfilled "I have my dream house in the city ... in the area... with or without a* *garden so on " be very specific about details: "It is a big house, with a bright hall at the entrance, with a large living room, an open kitchen and 3 bedrooms and 4 baths. Each bath has a shower" and so on, describe each room how you imagine it to be and, the color of the walls, the furniture, the windows, etc. Do not forget, **be very specific** and allow yourself to be carried out in the story of your house, see it be there, take walks in it, observe what emotions you feel now that it's yours! Live your dream as already fulfilled, see it in the present! You can meditate, dream on what you desire throughout the day as often as you can and depending on how much you want that "something". You can even set alarms on your mobile phone at various times so that it becomes a habit, until the proposed goal is achieved. And please, DREAM BIG!*

*Remember: the subconscious mind doesn't make any difference between fiction or imagination and reality so, **if you can see it in your mind, you'll definitely see it for real too**! The more intense the feeling the quicker the manifestation!*

Habits

Our habits are the repeated behaviors we engage in automatically, often unconsciously. Your task here is to be able to recognize those habits that support your objectives and make a conscious effort to reinforce and cultivate them. But you should also identify the harmful habits and the possible blocks and then replace them with positive and productive ones.

Suggestion:

Best way to identify recurring habits is by asking for professional assistance like psychological counselors, hypnotherapists, life-coaches or other specialists! The trap here is not to recognize your own habits that are blocking you or holding you back, therefore the best would be to ask a specialist but this is not possible, another way would be to ask your closest friends or relatives that you trust most.

Beliefs

Our beliefs are the thoughts and convictions we hold about ourselves, others and the world. It's crucial to examine your beliefs and assess if they serve your growth and success. Challenge any limiting beliefs and cultivate empowering ones that align with your objectives.

For example, when we feel uncomfortable about something we believe we don't or cannot do or like, we tend to run away by indulging in a compulsive behavior that gives us a dopamine spike like binge eating, scrolling through social media, watching Netflix! That's the natural impulse: escaping from hardship! BUT - Do not run away from your feelings, run through them! Identify the feelings you have, sit quietly with those feelings and get through them, becoming aware of them and the reasons you feel that way. It's like with "the fear of something", the best way to overcome it is to actually do the thing you are afraid of, go ahead and do it again and again and again until you almost feel numb to it! And you end up feeling that you're not afraid of it anymore, it doesn't affect you as much anymore!

Actions/Behaviors

Our actions determine our behavior and are the physical expressions of our thoughts, habits, beliefs, and values. To reach your objectives, break them down into specific, achievable actions. Take consistent and intentional steps towards your goals, even if they are small.

Remember, breaking habits and beliefs takes time and persistence. Be patient with yourself and celebrate even the smallest victories along the way to maintain motivation. By using the sabotaging behaviors technique and staying committed to your desired change, you can effectively break old patterns and develop new, empowering behaviors and beliefs.

Suggestion:

You know what they say: "Everybody sees Bill is just like his father but Bill". Well, this being said, let me share with you one effective NLP technique called "sabotaging behaviors." This technique's aim is to interrupt and eliminate the automatic patterns of behavior that reinforce unwanted habits or limiting beliefs. Here's how you can apply it:

First, identify the habit or belief: Start by pinpointing the specific habit or belief you want to change (ex. nail biting). Clearly define what it is and how it manifests in your life, write it down. Be specific about the behavior associated with it.

Secondly, be mindful and self-aware to recognize when undesired behavior or belief is triggered (when I am angry, or when I am anxious). Pay attention to the thoughts, emotions, and physical sensations that arise when engaging in behavior or contemplating the belief (observe objectively your thought and emotions).

Thirdly, create a pattern by developing a deliberate, unexpected action that interrupts the habitual behavior or thought process (take a chewing gum or start singing or you can ask someone to slap you when seeing you biting your nails). This should be something that surprises you and breaks the automatic nature of the behavior or belief. (Another example: you could snap a rubber band on your wrist, say a specific phrase aloud, or visualize a stop sign).

Fourthly, substitute with a new behavior or belief: immediately after the pattern interrupts, consciously substitute the old behavior or belief with a desirable and empowering one. This can be easily attained with the help of a specialist or learning the method and do it yourself by creating an anchor (* please see the description of this technique at the section allocated to the techniques). Engage in a different action or focus on a positive affirmation that counteracts the old habit or belief. Repeat and reinforce this substitution consistently.

(I am feeling better and better each day, feeling calm and released)

Fifth, visualize and reinforce: utilize the power of visualization to create mental imagery of yourself successfully engaging in the new behavior or holding the empowering belief. This visualization can enhance your motivation and belief in the change you want to make (use the same breathing technique described at the Suggestion point for listening to your intuition to better decide upon the sources).

Sixth, consistency and repetition: breaking habits and beliefs requires consistent effort. Set reminders or triggers throughout your day to reinforce the pattern interrupt and substitution. Repeat the process every time you catch yourself engaging in the old behavior or thought pattern.

Seventh, seek support when needed: if you're struggling with breaking a particular habit or belief (which can easily happen given the old habits and believes being sometimes very well impregnated into the subconscious mind) you can consider seeking support from a coach, therapist, or even an accountability partner. They can provide guidance, encouragement, and assist you in maintaining consistency and accountability.

Environment

Our environment includes the actual physical spaces we occupy, as well as the people and all the other influences around us. In order to attain our desired goal, we must first create an environment that supports our goals and fosters positivity.

It is equally important to surround ourselves with individuals who inspire and encourage us while eliminating the so-called toxic ones. Our physical space must be organized and favorable to productivity.

When we want something, there are always at least two possibilities and we can always change our minds. Therefore, we must first decide upon the exact thing or goal we want to achieve. The decision must be the final point and once taken, there is no way back.

To make better decisions and achieve our objectives we can:

1. Seek diverse and reliable sources of information and knowledge that we can trust.
2. Clarify and prioritize your values to guide your decision-making.
3. Cultivate positive habits that support your objectives.
4. Challenge and replace any limiting beliefs by creating new better ones.
5. Focus on taking consistent action towards your goals with trust, confidence and the assurance that it is already manifested.
6. Create an environment that fosters growth and positivity.

Remember, progress is an iterative process. **Embrace mistakes as learning opportunities** and **adjust your strategies accordingly. You are the creator of your own reality, you have the power to shape your future the way you like it to be**, so stay committed and resilient.

If you let doubts infiltrate, even for a bit, then delay must be faced and even worse, the opposite can be manifested. Cause you see, the universal law of cause-and-effect states clearly that for every cause, there is an effect (pretty straightforward)! Or similar to Newton's third law, "For every action, there is an equal and opposite reaction". If you move your hand through the air, you are causing molecules and the space around your hand to respond. And we know now that our actions are the mirrors of our thoughts and emotions so, dear ones if doubt is what you think about and feel, then guess what will you attract into manifestation, based on the law of attraction and the law of cause and effect? Nothing less, nothing more, but what you constantly think and feel! So, if you don't like what you see, now you know: it's nobody's fault but yours!

Put your thoughts in order, align with your goals and keep your trust till manifested!

Even though it's all arranged by the book, and we are all set to go for our goal, there are moments when our old habits kick in and destabilize us, leading to disappointment!

And then there are times when we seem to be going in circles and failing to mobilize and organize when we set out to do something new. There are those times when we find the best excuses along the way, or we get easily distracted by something that seems more interesting than focusing on our goal. Does it sound familiar?

Why do we prefer to procrastinate and bargain with ourselves when it comes to doing something we want, but it never seems like the right time yet? Why do we usually resist change, even though we want that change to take place in our lives? Well, the answer is because we are actually afraid of change and we try by any means to hide this, because we believe that "we are not able to do it" but we do not want to let this to be seen. It makes us feel insecure and triggers the belief that "we will not be good enough" or that "we are not deserving" and, so on. As follows, we lose control of what we know as "normality" and this destabilizes and blocks us at the same time.

Self-sabotaging is "comfortable" although it displeases and discourages us, for the simple fact that this is how programming is manifested which over time becomes a habit that turns into a normality, the so-called comfort zone. That "normality" in which it seems normal for us to suffer, to struggle in order to succeed, not to have luck, to turn everything upside down.

All these are behavioral patterns unconsciously appropriated from **parental and social programming.** Remember the old saying about Bill?

The way our parents encouraged us, supported us, and guided us or not to trust our decisions, to have the determination and enthusiasm to do that which seems impossible, to experience the mistake and its consequences and then to find the strength, inspiration and courage to start over and not to give up until we succeed, has had a direct impact on building self-confidence and self-esteem, as well as the habit of giving up because it seems too complicated to continue, no matter how challenging it is.

As for **social programming,** that comes as a continuation of **parental programming**, it "certifies" that what our parents have been trying to convey to us, remains valid and continues to contribute. How? Starting with the way they told us in school what we can do, but especially what we can't do. We are conditioned, manipulated, pointed at and punished when we make a mistake, we are discouraged from having too many dreams because the chances of success are limited and so on.

Then there are the media channels, which for the most part urge us, through the diversified content, to get lost in doing things that support our state of dissatisfaction, concern, insufficiency, distrust and fear.

Besides these two programs, the parental and the social ones, there is a third one, the toughest of them all. But here only we can have the freedom to choose: to continue our lives finding "nodes in the rush" every time or to enjoy life, learning from challenges and leaving the "comfort zone".

This third programing is **self-programming**. And in order to be able to get rid of the unconscious habit of self-sabotage, it is necessary to understand our emotional patterns and wounds that formed it, and later on, to work on the behavioral pattern that we automatically manifest when our comfort zone is endangered. So, what we must look here into are:

- Causes and effects of behavioral programs and patterns.
- Self-sabotaging habits, beliefs and fears; and
- The way they unconsciously influence our moment-to-moment choices and what we can do to free ourselves from all this and start living the life we want, to our true potential…

Self-awareness, as stipulated before, plays a crucial role in our quest to unveil our true identity. It requires a thorough understanding of our strengths as well as weaknesses, and patterns of behavior. By following the steps within the pyramidal figure described before, starting from the Source and ending with the Environment, we can realize many of these desiderates and start applying them in our daily lives. Only by a sustained and continuous introspection and reflection, we can identify the limiting beliefs and negative self-talk that may be holding us back. By replacing self-doubt with self-compassion and self-belief, we can unlock our untapped potential and step into our power. "Love thyself! "must be the starting point in one's journey to self-discovery and self–awareness, also the final step but from a broader perspective. It is like the painter that is born with his talent but doesn't become aware of it until he grows out of the studies and his life experiences and draws a simple line on a canvas that encompasses all his knowledge that he transmits almost telepathically to the public and gives birth to a masterpiece of art!

It is well said, long time ago, that there is nothing discovered under the sun but only rediscovered and so goes for the human being: imagine yourself being like a rose that blooms slowly and with each petal you unveil a new trait or aspect of yourself to only better understand your identity, who are you in this life, here and now at the certain location! Keep always in mind though that, as movement is the trait of the universe, the life force itself, you also are never the same. ***Every day you are a new you, but the secret here is to be every day a better version of yourself!***

One of the helping tools we can rely upon on your journey of self-discovery is life coaching that can be an invaluable instrument. A skilled coach can provide guidance, support, and accountability as you navigate the challenges and uncertainties along the way. It can help you uncover your unique strengths, clarify your goals, and develop strategies for personal growth. With the assistance of a coach, you will gain confidence to overcome obstacles and create a life that is aligned with your true identity. The important thing to remember here is to find the coach that resonates the best with you and you should not stop looking for the "one" until you find it! There is always there the proper master ready for you when you are ready for him – at the so called "right time" that brings forth the right people for you. But actually, the proper meaning of these sayings is that the right time and the right people are actually appearing in your life when you are ready or perfectly aligned!

In conclusion, uncovering your true identity is a transformative process that empowers you to become a powerful and complete being. By reflecting on your values, passions, and aspirations, cultivating self-awareness, and seeking guidance from a coach, you will embark on a journey of self-discovery that will enable you to navigate life's challenges with confidence and grace. Embrace your authentic Self and unleash your true potential – the world is waiting for you to shine.

Suggestion

Anchoring technique

"Anchoring" is one of the fundamental Neuro-Linguistic Programing (NLP) tools, which can be very powerful in helping you to have more confidence, enthusiasm and be more relaxed when meeting people or other circumstances.

It's a simple way that allows you to change an unwanted feeling into a resourceful one in a matter of minutes. When you create an "anchor" you set up a *stimulus response pattern* in order to feel the way you want to, whenever you need to.

In NLP, "anchoring" refers to the process of associating an internal response with some external or internal trigger so that the response may be quickly, and sometimes covertly, re-accessed. Anchoring is a process that on the surface is similar to the "conditioning" technique used by Pavlov to create a link between the hearing of a bell and salivation in dogs. By associating the sound of a bell with the act of giving food to his dogs, Pavlov found he could eventually just ring the bell and the dogs would start salivating, even though no food was given. In the behaviorist's stimulus-response conditioning formula, however, the stimulus is always an environmental signal and the response is always a specific behavioral action. The association is considered reflexive and not a matter of choice.

In NLP this type of associative conditioning has been expanded to include links between other aspects of experience than purely environment signals and behavioral responses. A remembered picture may become an anchor for a particular internal feeling, for instance. A voice tone may become an anchor for a state of excitement or confidence. A person may consciously choose to establish and re-trigger these associations for himself. Rather than being a mindless knee-jerk reflex, an anchor becomes a tool for self-empowerment. Anchoring can be a very useful tool for helping to establish and reactivate the mental processes associated with creativity, learning, concentration and other important resources.

It is significant that the metaphor of an "anchor" is used in NLP terminology. The anchor of a ship or boat is attached, by the members of the ship's crew, to some stable point in order to hold the ship in a certain area and keep it from floating away. The implication of this is that the cue, which serves as a psychological "anchor" is not so much a mechanical stimulus, which "causes" a response, as it is a reference point that helps to stabilize a particular state. To fully extend the analogy, a ship could be considered the focus of our consciousness on the ocean of experience. Anchors serve as reference points, helping us find a particular location on this experiential ocean, holding our attention there and keeping it from drifting.

Imagine what it would be like if you could, in a moment, go from feeling anxious to feeling decisive and absolutely capable right in the middle of a stressful interview when all eyes are on you, or dealing with an individual you struggle to get along with.

Step 2. Embracing Your Strengths and Weaknesses

In our world that often emphasizes perfection and flawlessness, it is essential for human beings to embrace their strengths and weaknesses. Recognizing and accepting these aspects of ourselves is a fundamental step towards personal growth and empowerment. In this subchapter, we will explore the importance of self-awareness and how it can lead to a powerful and fulfilling life.

Self-improvement through self-awareness is the core of becoming an empowered being. And it begins with taking the time to reflect on who you are as an individual:

- Identify your strengths – those unique qualities and talents that make you shine. Embrace these strengths, celebrate them, and use them to your advantage. By acknowledging your strengths, you can build upon them, honing your abilities and becoming even more powerful.
- On the other hand, understanding your weaknesses is equally crucial. We all have areas where we may struggle or

feel less confident. Embrace these weaknesses as opportunities for growth and learning. By recognizing them, you can develop strategies to overcome them or seek support from others who excel in those areas. Remember, weaknesses do not define you; they are merely areas where you have room to improve.

You can tap into the mysteries of Numerology or Mind mapping techniques and get professional help to find your path life, your soul urge or personality traits, positive and negative aspects and how to better navigate the sea of life. Or you can study yourself how to calculate all that and you can get one of my courses for this purpose (you can find all the details at the end of the book). Learning to calculate and understand numerology profiles will help you comprehend your Self better but also how to improve your life in all aspects: personal, relational and financial.

If you are interested in knowing more about these techniques, you can always take my course: to learn about the importance of numerology, about how it shapes our lives and existence and about how to recognize its influence on your life.

But first of all, we must dig a bit into the depths of this science, because yes, numerology is a science and you should first understand what it is about and how can numbers actually influence our lives? Well, the concept of Numerology is said to have originated from the civilizations as old as Babylonians (current day part of Iraq), Egypt and even older. The ancient civilizations believed that numbers had true powers and that they comprise the essence of the divine, which led humans to know more about life and God. It is during this time that the Chaldean System of Numerology was derived and practiced. There is also enough corroboration to show that different forms of Numerology were also practiced in China, Rome and Japan as well, which proves that people have been following this idea and notion of the relationship between numbers and events from a very long time. The Chaldean alphabet's number numerology system was itself inspired by the Indian Vedic numerology. Several types of Numerology were practiced, depending on when they were discovered and how they were used, but the below-mentioned ones are the most predominant.

- Chaldean
- Kabbalah
- Vedic
- Western/Pythagorean

Chaldean Numerology is one of the oldest forms of Numerology that was discovered in Ancient Babylon. Unlike the Pythagorean system where alphabets determine the calculations, this form of Numerology is based on the vibrations that each letter emits, which in turn, helps to further calculations. The digit numbers assigned for alphabet's letters are 1-8 not 1-9 like in the Pythagorean's system.

When discussing types of numerology, Cheiro's contribution to Numerology cannot be ignored. Cheiro, in the last century, popularized the Numerology system to the form that we are aware of now. Cheiro's original name was William John Warner.

The numerology adopted by Cheiro had a few accurate subjects, but he pays so much attention to minor details, like the number of individual letters in a name or adding random numbers to somehow correlate to a number value, which makes it very hard for a layman to believe in this system. Many events such as the Russian Revolution and its alliance with China, World War II, the 1926 British Trade Union Strikes, developments in the Middle East, and numerous historical events in the UK royal family were all predicted by Cheiro and several of them came true during his lifetime.

Cheiro used Chaldean numerology, which is less widely used today compared to Pythagorean numerology. In India, the majority of numerologists use Cheiro, while Pythagorean one is commonly used in Western countries, and hence, it's very popularly known as Western Numerology.

The Kabbalah Numerology is associated with the Jewish culture where the name of the person is taken into consideration in the process of calculations rather than the date of birth, which does not hold much importance as far as the Kabbalah system of Numerology, is concerned.

The Vedic Numerology originated from the southern part of India (Tamil), is also known as the Indian Numerology. This form of Numerology takes into consideration three essential aspects about each person while the calculation is made. These are the Psychic Number, the Destiny Number, and the Name Number.

Western Numerology, also known as Pythagorean Numerology, was discovered by none other than Pythagoras himself. In this system, each letter of the alphabet is assigned a specific numerical value, which further helps in the process of calculation. Modern Numerology is based on the Western/Pythagorean Numerology. Pythagoras is considered to be the father of Numerology in the modern day because of the amount of knowledge that he had gathered and shared. According to Pythagoras, numbers play an important role in identifying the nature of anything. Furthermore, as he was also more inclined towards mathematics, he considered numbers to be an authentic and a practical source to know the absolute truth. However, in spite of this, people were still not aware of the magic that one can create with numbers, but the church was always against it. Although some skeptics and non-believers found it difficult to accept the fact that there can be a rational explanation behind this notion, Numerology saw a boom in the latter part of the year. Moreover, up till now, a lot of research has been done which can easily prove that numbers play a vital role in everyone's life and that each number has a specific meaning.

Numerology has now become a very essential part of our existence, in knowing more about one's life, the personality that they carry and in knowing the relationship between the numbers and the incidents that have either taken place or may take place in the future. Numerology in the modern world has proved to be a sufficient tool to know more about life and its mysteries along with comprehending the absolute truth of life, death and the universe itself!

Pythagorean numerology is a device that enables you to create a full Numeroscope, depicting all of the person's characteristics from the inside out and delineating the results in detail on every level, as well as deciphering a clear-cut trend/pattern within the numbers.

The modern numerology assigns a distinct value to each letter of the alphabet, which is similar to the Pythagorean Numerology. For example, A has a value of 1. B has a value of 2, and so on. Therefore, a lot can be derived from a single name of a person, which can help in determining the quality of his life, his personality and much more. In such a case where a name has to be converted into numbers, each number of the corresponding letter of the alphabet from the name has to be added until you reach a single digit number by the process of Fadic addition. This means that all the values allocated to each alphabet have to be added repeatedly to reach a single digit number.

Chaldean and Pythagorean numerology

While each one of the 3 pillars of Numerology, namely Pythagorean, Chaldean, and Kabbalah, have their own set of benefits and drawbacks, deciding which is the best depends on your goal. Let us compare in brief Chaldean and Pythagorean numerology analysis to understand what is most suitable for you. Pythagorean numerology system is very convenient for a new student of Numerology.

If you are someone who has spent significant time doing Numerology, and you want more accuracy, then the Chaldean Numerology system might be suitable for you. However, when you are strolling through uncharted territories, it is sensible to consult a guide, so consult an experienced Numerologist to make a choice between the Chaldean and Pythagorean numerology analysis.

It needs some effort on your part to understand which of the numerology system best suits you. Whichever makes sense to you or appeals to you is perhaps the most one suitable for you. But no matter which numerology system you choose to go ahead with, they all have their own designs on which calculations are based, and looking at this, it can be said that every numerology system is unique. Those will simply produce effective results based on integrity and faith. Depending on their capabilities, various countries can use different numerology systems.

Pythagorean numerology analysis, during recent times, has gained immense popularity, especially amongst celebrities. This method may have been less common in the past, but according to some new-age numerologists, this is nowadays the most accurate numerology system. Digital media is full of different types of online calculators, and you will find many numerologists online claiming to be the best. You must not hesitate to discuss their understanding of all types of Numerology before you proceed further. A numerologist is not just a student of this science and whatever course one takes makes him an expert, since the complexity of it encompasses knowledge of many other disciplines as mathematics: geometry (mostly the sacred geometry), physics, philosophy, astrology, psychology, sociology etc. and most important, the experience of life! It was not for pure words that the students of Pythagoras were swearing an oath and never releasing details of their studies with the Master!

The worldly well-known dictate of Pythagoras's school was "All is number" or "God is number", and the Pythagoreans secretly practiced a kind of numerology or number-worship and considered each number to have its own character and meaning. They were following very specific rules like a specific diet based on plants but no favas since Pythagoras considered them seeds or ancestors' souls that incarnated in these so, there was no way to eat them! Now, if someone wants to discover more about Pythagoras, there are many stories and myths about his origins, some claim he is the son of God Apollo or even Apollo himself, while others consider him the son of the very one Hermes Trismegistus, henceforth his knowledge!

Pythagoras discovered **prime numbers and composite numbers** (any integer that is not a prime). He also looked at perfect numbers, the ones that are the sum of their divisors (excluding the number itself). For example, six is a perfect number; its divisors are 3, 2 and 1, and 3+2+1=6. Pythagoras' secret symbol was the pentagram, also known as the pentacle. It was considered a symbol of the Pythagorean brotherhood and was associated with the mystical properties of the number 5 (five).

In module 2 you can already start calculating the life path or destiny by simply adding your date, month and year of birth until you reach a single digit. Exceptions are the cases in which the total number is made of the same two repeating digits like 11, 22, 33, 44, these are named as Master Numbers. These Master Numbers (MN) aren't meant to forge your life path like doubling it, due to a repeated digit; instead, they're meant to complement your experiences. If you're not sure why you responded the way you did in a given situation, the interpretation of the energies of these numbers can help.

Ex. If someone is born 02.08.1976 than his life path will be:

- 0+2+0+8+1+9+7+6= 33, in this case we can further add 3+3=6 and his life path could be 6 but since the final sum is 33, this is a Master Number and the mission of the person is above the one of the number 6. In Numerology, the numbers 11, 22, and 33 are so exceptional that they've been dubbed as "Master Numbers." These profound energies, when combined, embody the three stages of creation: envisioning, creating, and sharing. The 11- the Seer is the also called the teacher, 22- Master builder, is the planner, the visionary, and 33 - Master teacher, is the messenger the combination of the previous two. More about these numbers you can find out reading the life paths symbolism.

- For the simple digits life paths 1-9 you find below the explanations; for ex: if someone is born 05.11.1989 his life path is 0+5+1+1+1+9+8+9= 34, and 3+4=7, so 7 is the destiny or life path.

Life paths meanings

Life path 1 – the trailblazer, the pioneer. Number 1 symbolizes the masculine principle of creation – the beginning, the sharp blade that depicts the matter in two. It is the initiator, the first line in geometry –the horizontal line! The One, the unique! The Hero who starts his journey!

You are a purposeful and goal-oriented person who has a lot of determination and dedication to reach your goal. You have the power to achieve what you want in life, with hard work and resolution. You put your heart and soul into accomplishing your goals. You are always willing to take on problems and challenges that come your way. You are very clear in your path and quite responsibly take up hurdles to meet your needs. Furthermore, you as life path 1 in numerology need attention as well as affection from people around you and in return, you safeguard them and treat them with love and respect. Sometimes you may become a bit irritated and agitated when things are not going your way or as planned, which has a negative impact on you.

And so on....

If you are interested in knowing more you can get one of my courses in numerology.

It is important to remember that embracing your strengths and weaknesses is not about comparing yourself to others or striving for perfection. It is about recognizing your own unique qualities and accepting yourself as a whole. Embracing your strengths and weaknesses allows you to live authentically, with confidence and grace. It is about taking your power, getting to live by assuming responsibility for your decisions where fears are identified, integrated and dissolved into courage that gives you the real strength to create your own life the way you would like it to be. Like this, you stop the chain of being a victim of the system, of God or some other supernatural force and you become the HUMAN BEING you were born to be here and now.

As you embark on your journey towards becoming a powerful and complete being, remember that both your strengths and weaknesses are what make you beautifully human. Embrace them, nurture them, and continue to grow. With self-awareness and a commitment to personal growth, you have the power to create a wonderful life. Mostly we are hearing about getting rid of our own Ego whereas my personal opinion is that this is a totally wrong approach: rather conquer you Ego, embrace it and love it, fully understand it by:

- Inward reflection
- Becoming aware and then integrating your negative aspects through love: remember that we are love.

Giving up to Ego is synonym to death. Who am I without my Ego? Some viewpoints are suggesting to "better put your ego to work for you not against you" and you'll have a better existence.

People from beautiful Bali believe that in order to stay happy one must always know exactly where he is, in every moment! "The perfect place to be is right where sky meets the earth! Not too much God, not too much Ego or selfishness!" You must be, as much as possible, at the Equilibrium point.

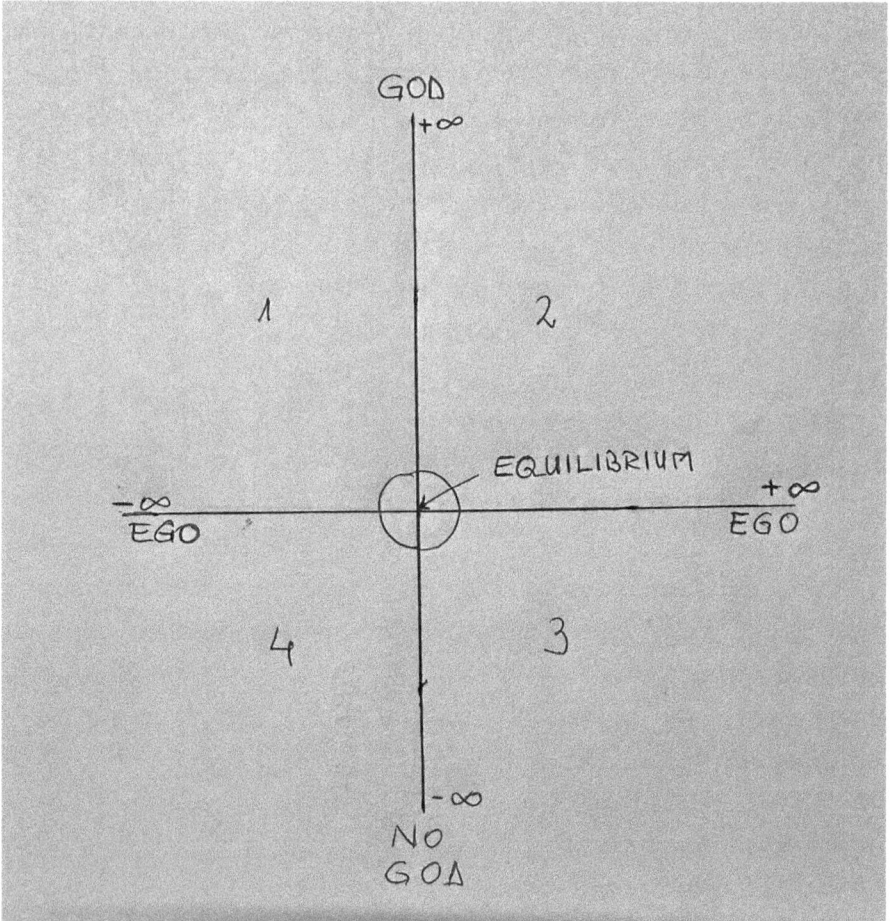

Keeping yourself in the balance point is the secret of a harmonious life otherwise, all is too chaotic! When you lose your balance, you lose your power! You have to learn to select your thoughts same way you select your clothes every day! That's the power you can cultivate!

Stop trying, surrender!

If you want to control your life, then educate your mind! Learn to manage your thoughts! But for that to happen dedication, steadiness and commitment is required! So, start your new day with a good smile, be grateful for the fact that you are here, alive and meditate everyday morning and night about you and all your dreams! Fall in love with yourself! You deserve the best you can imagine! And there is plenty of it in the whole Universe and Multiverses!! If only you would allow it yourself! I promise you that the moment you give up your old paradigm, a door will open and the entire Universe that cannot wait anymore, will pour all the BEST into YOU! Understand, beautiful soul, that **you are loved, beyond words and God wants you to be happy and to live your dreams' life, NOW!** God made us a promise when we were born, to each and every one of us, and that is LOVE! But us, entrapped in all the societal meanders, we forgot about this most important promise of our life: I Love You, no matter what!

So, sing, dance, eat, pray, meditate, cry, be mad so you can be calm again just simply allow life to unfold for you without fears, doubts or judgments that limit your Soul's journey! Life is a wonderful voyage if you know how to enjoy it! It's up to you! But remember that you, the real you as you are now, has just this time, this life to live as You and never again!

So, how would you like to think about your life when you'll be old? Would like to tell a sad story or a happy one? Take a deep breath and meditate about it!!! The meditation room is within you! Make your choice!

Remember: God loves you and wants you to be happy but He will never force you, because He loves you and respects your choices, no matter what! But please, do not be like the poor man in the story, who went every day to pray to a saint statue and icon:

Oh, my Lord, please, please, please help me win the lottery! After a while the saint came to life and said to the man: my child please, please, please buy a ticket!

Step 3. Cultivating Self-Compassion

In today's fast-paced and demanding world, it is easy to get caught up in the pursuit of external success and validation. However, true empowerment and fulfillment comes from within. No matter how much wealth, success or fame one acquires, if there is no balance attained at the inner level, if the person is not happy, satisfied with herself, all the rest are simply not enough and will never be enough by themselves! And there are so many examples in the world: the lottery winners that lose their money as quickly as they won it, famous persons like actors, businessman who are depressed and end up even to commit suicide!!

In our pursuit of becoming powerful and complete, it is crucial to cultivate self-compassion as a foundation for personal growth and transformation. So, let us delve into the importance of embracing self-compassion and how to use practical strategies for cultivating it in our lives.

Self-compassion is the practice of extending kindness, understanding, and support to us, especially during challenging times. It is about treating ourselves with the same love and compassion we would offer to a dear friend. Cultivating self-compassion allows us to embrace our imperfections, forgive ourselves for mistakes, and develop resilience in the face of adversity.

To begin, it is essential to develop self-awareness. Understanding our thoughts, emotions, and patterns of self-criticism is the first step towards self-compassion. By observing our inner dialogue and identifying negative self-talk, we can consciously replace it with positive and affirming thoughts. This process requires patience and a commitment to self-improvement.

In addition to self-awareness, self-care practices play a vital role in cultivating self-compassion. Taking time for ourselves, engaging in activities that bring us joy, and prioritizing our wellbeing are essential aspects of self-compassion. Whether it is practicing mindfulness, engaging in physical exercise, or indulging in creative pursuits, these self-care practices nourish our souls and help us build a stronger sense of self.

Another powerful tool for cultivating self-compassion is practicing gratitude. Taking a moment each day to acknowledge and appreciate the positive aspects of our lives fosters a sense of contentment and self-acceptance. Gratitude reminds us of our strengths, accomplishments, and the abundance that surrounds us, helping us to develop a kinder and more compassionate relationship with ourselves.

Furthermore, surrounding ourselves with a supportive community is crucial in our journey towards self-compassion. Engaging in meaningful connections with like-minded individuals, seeking guidance from mentors, or working with a life coach can provide invaluable support and encouragement.

By cultivating self-compassion, we can navigate life's challenges with confidence and grace. This subchapter serves as a guide to help every person who wants to embrace self-compassion as a transformative practice. Through self-improvement and self-awareness, we can create a life of empowerment, fulfillment, and genuine success. Remember, you deserve love, compassion, and kindness - from others and most importantly, from yourself.

Suggestion

A strategy to follow to start your day with a 5-minute meditation. Just after you wake up, keep your eyes closed and repeat in your mind "what a wonderful being I am" and as you say that see yourself smiling, feel your heart feeling with love and gratitude! Enjoy being

YOU! And embrace being YOU every day with the positive and the negatives and pay attention to your thoughts and feelings! Observe yourself – stay as much as you can in balance – keep neutral –the Observer of the observer! Remember what Buddhists say that all is Maya, an illusion!

Step 4. Establishing Boundaries for Self-Preservation

In the journey towards empowerment and self-fulfillment, one of the most crucial aspects is establishing boundaries for self-preservation. As human beings seek to become powerful and complete, it is essential to recognize that our wellbeing and happiness depend on our ability to set healthy boundaries in all areas of our lives.

Boundaries are like invisible fences that protect our physical, emotional, and mental space. They enable us to define what is acceptable and what is not, allowing us to take control of our lives and protect ourselves from negativity and toxicity. By establishing boundaries, we create a safe and nurturing environment in which we can thrive and grow.

Self-improvement through self-awareness and coaching plays a significant role in the process of boundary setting. It always begins with the deep understanding of who we are, what our values and priorities are, and what brings us joy and fulfillment. By connecting with our inner selves, we gain clarity and confidence, which are essential in establishing and maintaining boundaries.

Setting boundaries involves both external and internal aspects. Externally, it means learning to say "no" when necessary, asserting ourselves, and communicating our needs and limits to others effectively. It may involve setting limits on work hours, personal space, relationships, and even social commitments. By doing so, we create a space where our time, energy, and emotions are respected and protected. I know that so many of you are finding it so hard to say "no" especially to persons you care about! But no matter how difficult this might be for you, the most efficient way is to do it, purely and simply like that. Set your boundaries with the eye of your mind and repeat the phrase while visualizing the limits and the borders of your space: *I am the guardian of my personal space and territory, and I let in just the ones I choose whenever I choose! I am the master and I decide when I say NO and to whom I say NO!* Repetition will enforce it into a habit, a good one!

Internally, establishing boundaries requires us to recognize and challenge our own self-limiting beliefs and behaviors. It means letting go of toxic patterns, negative self-talk, and perfectionism. By acknowledging our worthiness and deservingness, we can embrace self-care and self-compassion, which are essential for our overall wellbeing.

Establishing boundaries for self-preservation is not always easy. It may require courage, assertiveness, and the willingness to confront uncomfortable situations. However, the rewards are immeasurable. By setting boundaries, we reclaim our power, restore balance in our lives, and create space for personal growth and fulfillment.

As human beings seeking empowerment and completeness, let us remember that establishing boundaries is not selfish; it is an act of self-love and self-respect. It allows us to be our authentic selves, to live life on our terms, and to surround ourselves with people and experiences that align with our values and aspirations.

In conclusion, our journey leads us towards becoming empowered begins with the intentional establishment of boundaries for self-preservation. Through self-improvement and self-awareness, we can learn to set healthy boundaries, both externally and internally. By doing so, we create a nurturing environment that supports our growth, happiness, and overall wellbeing. By having the courage to set boundaries, we are taking a powerful step towards a transformative path of becoming the best versions of ourselves!

CHAPTER 3

———— ·◆· ◆ ·◆· ————

Building Confidence: Unleashing Your Inner Strength

A Philosophical Debate

When we are considering the two existential questions, that are occupying the first top seats on the philosophical debates: *who or what we are* and *what is it with us here on Earth*, I would like to dedicate a special attention to different culture's belief or different currents that become more and more trendy nowadays and are adopted by the modern world, so we better understand the approach holistically.

There are many starting points I can choose from but the one that pops into my mind is another famous question: *does life happen to us or we just happened to live life?* And the answer should depend on each individual's choice even if made unconsciously. Many people may leave their life almost unaware of what is happening externally, around them and more importantly, internally whereas others, fewer indeed, are more consciously aware. So, what makes us choose how to live our life? I believe everyone or at least most of us should be interested in being happy and healthy and wealthy and the list can go on. All the religious texts of all religions of the world and all the philosophic currents or scientific studies are easily available nowadays to almost each one of us to research, study and observe, if we want. ***Truth is hidden in plain sight!***

Mind and thinking - Love and strength

Buddhism is one of the philosophies more than a religion that is oriented all about the "you", nothing more, nothing less! The main goal in Buddhism is to understand the true nature behind your life and your mind. Mind is considered the most powerful thing in the world. Buddha said, ***"The mind is everything. What you think you become!"*** which reminds me of the first Hermetic law, the Mentalism: *"All is mind!"*

In Buddhism there is a say: "If you get shot by an arrow do not ask where it came from nor who shot it, your own concern should be how to get out this arrow!" Because that is really what you should focus on, under the circumstances! Upon Buddhism believes, we live in the moment, we don't own our past or our future, we cannot control either of them, the only moment we have is NOW! The "arrow" in life is actually the "suffering" and the root cause or reason of why we suffer is because there is attachment in our mind; we get attached to people, we get attached to things, to the experiences.... And once we manage to let go of the attachment, we'll be happy! You may ask yourself why attachment is not good, why is attachment causing sufferance, it's good to be attached to your girlfriend, to your boyfriend, to your husband or wife, to parents, to your kids? Attachment is happiness, why not to be good? Well, the idea is that when we get attached to things, to people we have expectations and when some of our expectations don't get fulfilled, we become sad and automatically we suffer, ***expectation is the thief of joy***! Even if you suppose that something would always be there it is not beneficial to be attached to anything! Because nothing lasts forever, everything is subjected to change, that's the truth of this world! The only thing that is not submitting to change is the process of changing! Buddha said: *"nothing is forever except for change."*

For many people may be confusing this concepts but, love and attachment are two different things: when you love someone, you do all in your power to see them grow and become happy but when you are attached to someone, you expect him or her to make you happy, wondering why they don't make you happy! Whereas we should love our partner, parents, kids, without expecting anything from them, without expecting them to make us happy! On the contrary we must remain focused, showing our care and affection to them. We should be like a tree, a big one: the birds come to eat the fruits and they are happy but then they go and we remain the same tree, tall and profound and free, without trying to stop the birds in any way! Dalai Lama once said that Buddhism is not to become a better Buddhist but to try to use it to become the better version of what you already are!

All these concepts are undeniably teaching us to remain centered and express love without any interest, just focus on giving and stop worrying about receiving!

Jesus was also teaching about love and in John 15:13 (NIV), he says: *"Greater love has no one than this: to lay down one's life for one's friends."* The Christian biblical definition of love is a selfless, sacrificial, and unconditional commitment to the wellbeing of others. Jesus's primary teachings about love were for everyone to become like the good Samaritan, to love and help everyone, even his or her enemies. This kind of love is defined as *AGAPE* love. It is God's love for all humans: the highest form of love, sacrificial and unconditional. This is not at all the kind of love the rest of us may think of when we relate to the good- feeling emotion we feel towards our parents, our spouses and children and friends! Nor that we feel towards our pets or clothes or even TV shows and Netflix movies, none of that is what Jesus means by love. The love He speaks about, *AGAPE,* is so radical even to us, in modern days and most probably it might have remained unfathomable to those who were there and heard him speak. Almost all his teachings in the Biblical Gospels are related to this radical form of love and the reason for this will become clear only when we unveil the primary role of consciousness in the construct of reality. This has to do with the understanding of the radical and transformational role that real love has once you make it the primary focus of your life! It becomes a magnetic force that drives your life and you become it, like all great masters and philosophers once said and understood it is the essence of our strength, the essence of life and evolution! And you cannot separate it from consciousness, it is the intrinsic tool of creation itself! How can life, existence of all that is, be without love?! A rhetorical question, because we simply cannot comprehend the wholeness of AGAPE in the making of the creation itself through our mere senses, unless we are willing and ready to transform and transmute our understanding into the consciousness! Like a newborn baby who simply abandons himself to the care of his parents without any resistance of any kind! So, are you ready to take this kind of journey? For this is a journey within, whenever we talk about love and this is the kind of love that Jesus talks about, the very one that leads to the discovery of SELF and Self-mastery! And this becomes your strength that you can use to move mountains! Once you decide! Remember, you always have the choice! But once you take the decision there is no way back no second thoughts: you become the force that pulls YOU, like a red thread, to your true path!

On the other hand, upon the more pragmatic Descartes, who's view became for a long time with his "I think therefore I am" more of a slogan for the human race, if only we are capable of thought because we can reason and express ideas, creative concepts then what about those who can't? A Harvard study, made some years ago, found that our mind is distracted on average about 47% of the time. By distraction, they meant that the mind was not where we expected or wanted it to be. So, as an example imagine you are reading this book when you start to think about the fact that your house is not so appealing to you anymore and maybe you need to redecorate it or sell it. Well, actually you wanted to stay focused on reading but your mind just took you away from it, without your consent, or even without your conscious awareness. Therefore, we may rephrase Descartes like: "I think, therefore I get distracted" or "I think sometimes but most of the time my mind does what it wants to do, whether I like it or not. I'm not at all in control of it."

The truth is that we are conditioned to get distracted: *"we are far more conditioned than we think we are and, ironically, we do not think we are conditioned precisely because we are conditioned not to think that we are conditioned!!!"* And this is actually the absence of awareness, of mindfulness! It's like when you drive your car and talk with another passenger, being actually totally absent minded about the road, and yet, you're the one driving!

This happens when we are not quite present in our lives because we are too concerned about our future and too preoccupied about what happened in the past! We are being dragged continuously by thoughts and emotions rather than us being in control of what the mind thinks, feels and chooses in any circumstance. Just think about it right now. Who gets angry when he gets angry; who chooses to feel depressed or anxious or hateful? Who decides that they'll carry a grudge around with them for 20 years or more? Who chooses to be something he denies he is? Well, dearest, it's not you or at least not the reasoning, intelligent, thoughtful you that Descartes suggests, because no reasoning, intelligent person would ever allow these toxic moods or views, anywhere near their precious mind! But imagine that he was right, that we really could be capable of fully controlling our thoughts, our feelings, moods, emotions and gut reactions. Under this circumstances I believe that what most of us will choose to feel would be love, happiness, joy, health, a sense of purpose and meaning, wellbeing. Most of us would also like everyone else to feel the same way so that a decent, secure, happy and purposeful society would develop.

And I cannot forget to mention my dearest mentor and most well-known author, Bob Proctor, whom I sincerely miss a lot, who dedicated his entire life to the profound study of the human mind and whom many consider him the world's foremost expert on the human mind. I will not have enough pages to write about his studies for the man lived and dedicated his life this therefore I will remain brief but, for the one who want more please read his books, he left a huge legacy to this world! As he always stated, even if he studied thousands of book, he followed not only the guidelines but he become a daily student, dedicating over 60 years of his life to the study of Napoleon Hill's book "Think and grow rich" that became his Bible, he never separated himself from it! Now, this is a dedication that came after a precise decision he took when he was at a very young age, the moment his mentor asked him to do everything in his power to become rich and the best version of himself! Back to our subject, Bob was so thoroughly explaining that our brains are simply run by the two minds: the conscious and the subconscious mind. He explained that our subconscious mind doesn't make the difference between what is real and what is imagined and runs the programs that we acquire and build along our lives defined as our paradigm! This paradigm actually dictates our reality or more accurately what we perceive as reality but may very well be an illusion! He postulated that we could change our reality once we change our paradigm! But it takes a decision! The difference between "decide" and "want" is made very clear: when you decide something there is nothing that can turn you back, but when you want something, you will always be faced with two options:

1. I want or
2. I don't want that something, leading you to renounce or have second thoughts!

Leaders take decisions quickly and stick to them, assuming their full responsibility! So, who do you what to be: do you want to be the leader of your life or do you want to think about it.

I love you Bob, you'll always exist in my heart, in my memories, and I will always honor you!

And since I mentioned, did you know that the latest discoveries proved that the heart has a memory of its own? The article appeared on the website of National Library of Medicine stated that: *"There are approximately 40,000 cells present in the heart known as sensory neurites which play a vital role in memory transfer. The heart is quite a mysterious organ, which functions as a blood-pumping machine and an endocrine gland, as well as possessing a nervous system. There are multiple factors that affect this heart ecosystem, and they directly affect our decision-making capabilities. These interlinked relationships hint toward the sensory neurites which modulate cognition and mood regulation."*

A very interesting research finding has been that the heart is involved in the processing and decoding of intuitive information (McCraty, Atkinson & Bradley, 2004). Previous data suggests that the heart's field was directly involved in intuitive perception, through its coupling to an energetic information field 3 outside the bounds of space and time (Childre & McCraty, 2001). Using a rigorous experimental design, there was evidence that both the heart and brain receive and respond to information about a future event before the event actually happens. Even more surprising was that the heart appeared to receive this intuitive information before the brain (McCraty, Atkinson & Bradley, 2004).

For decades, personality changes have been reported following heart transplantation, which include accounts of recipients acquiring the personality characteristics of their donor. Many categories of personality changes were observed like changes in preferences, alterations in emotions/temperament, modifications of identity, and memories from the donor's life. The acquisition of donor personality characteristics by recipients following heart transplantation is hypothesized to occur via the transfer of cellular memory, like epigenetic memory, DNA memory, RNA memory, and even protein memory. Other possibilities, such as the transfer of memory via intra-cardiac neurological memory and energetic memory, were studied as well. The future of heart transplants and its implications are further explored, including the importance of reexamining our current definition of death, studying how the transfer of memories might affect the integration of a donated heart, determining whether memories can be transferred via the transplantation of other organs, and investigating which types of information can be transferred via heart transplantation.

Now going back to the philosophical debate regarding the mind – our thoughts and the heart - as the place of our feelings, and memories, as seen above, Buddha got it so much more accurately than Descartes when he stated, "with our thoughts we create the world". That's so profound but at the same time it's also a bit frightening because when we are not in control of our thoughts the world we create can become a horrible one!

So, if you want a happy life for yourself remember to get rid of worries, anxiety, depression, anger, resentment, bitterness and all the other harmful and often self-destructive moods and emotions! You definitely have to train your mind to become less automatic, less conditioned, less distracted and, more under your direct control. This requires you to stay in the present, focused and to pay attention to your thoughts! If you want a life full of love, which is simply a state of mind, you have to develop it and if you want peace of mind then do the work that brings you closer to it, do not accept to be programmed and a puppet! If you envision a better world and would like to live in that kind of world then start with YOU by becoming the better version of yourself every day, a little bit more! And if you want to see change in others start by changing yourself, regarding others and the world around you, as Gandhi said, *"be the change you want to see in this world"*. And *"train yourself, so that you can become the change you want to see in yourself"*. In other words, *"first cast out the beam out of thine own eye; and then shalt though see clearly to cast out the mote out of thy brother's eye".*

And when the other may challenge you to get out of your good mood, or good feeling remember to breathe instead of jumping to react! Follow Jesus's teachings, when he said to turn your other cheek if you get slapped on one cheek! The meaning of this is not to literally act like this but instead to turn your head towards a new direction, a new perspective and remember that you always have the choice: to react and start a yo-yo action-reaction episode that can escalade or "to take your horses" to a totally different direction and keep your good mood like this showing that you are in control of your mind, your thoughts and emotions, and your reactions! Having no reaction is also an action! And in many cases, especially like this one, it's advised and proven to be the best one to follow! Remember, you always have a choice!

Among the Essenes teachings are the so called "Essenes mirrors" telling us that whatever we may not like or it upsets us about the others, it is either inside of us, meaning we have that trait inside of us but we are not consciously aware of it or, we are just judging it! Whatever the answer, the reason that we are faced with this is exactly because we must become aware of that specific trait so that by conscious reasoning and analytic understanding, we release ourselves, we set ourselves free from it.

Returning to the philosophy of the Buddhism, change in life is inevitable and our thoughts are constantly changing, as everything else around us. Friends and family can come and go and so can our belongings, therefore the more we try and hold on to things, the more grief and suffering this can cause us. What we have today come from the thoughts of yesterday but if we want to have a different future then we should better start thinking in the present tense about what we would like our future to be or see around us!

"To every problem there are three solutions: accept it, change it, or leave it! If you cannot accept it, change it. If you cannot change it then leave it". – Buddha

In conclusion, mastering your thoughts becomes a force you can use to mold your life as you please once you become aware of it.

Quantum discoveries proving our power

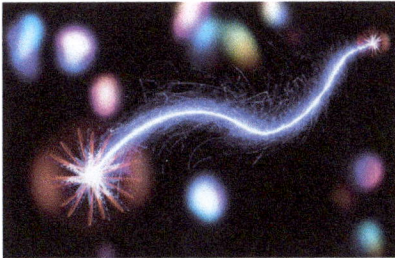

There is an undeniable force beyond our human understanding that nowadays, luckily, science comes and proves it. In recent years, the perspective provided by quantum information illuminated some of the most profound questions in physics.

For example, quantum information is the key to understanding the mystery of *black holes* and most probably of the entire universe. It also leads us to new quantum technologies that quickly and automatically encode and process quantum information.

But what is this quantum entanglement theory saying and why should that interest us or, what would we benefit from it? Well, in the simplest terms, it states that aspects of one particle of an entangled pair depend on aspects of the other particle, no matter how far apart they are or what lies between them. These particles could be, for example, electrons or photons, and an aspect could be the state it is in, such as whether it is "spinning" in one direction or another. The odd part of quantum entanglement is that when you measure something about one particle in an entangled pair, you immediately know something about the other particle, even if they are millions of light years apart. This odd connection between the two particles is instantaneous, seemingly breaking a fundamental law of the universe.

As stipulated in an American scientific article, Albert Einstein famously dismissed it as *"spooky action at a distance. Calling entanglement spooky completely misrepresents how it actually works and hinders our ability to make sense of it. Through the lens of quantum information, then, entanglement is not strange or rare, but rather expected. Einstein wanted all of nature to be identified with a simple and compact classical description. But we now know that quantum information provides the most accurate description of nature, which is written in a language we do not speak"*. There is a force that unifies, governs all that is or, in order words, we should remember here the famous say: nothing happens without a reason!

The *Unified field* or the Matrix attempts to tie up all known fundamental forces and the interactions between them in a single theoretical framework. The potential for such a theory was seen when Maxwell[3] tied up the electric and magnetic forces together into a new formulation of electromagnetic fields...

In his groundbreaking book, "The Scientific Proof of God", Fredrick Swaroop Honig takes difficult concepts about quantum physics and the functioning of the universe and explains them in simple language that anyone can understand. He integrates science and consciousness and shows us how this information contains the very core of our purpose on this planet. I invite you to read it if there is any interest manifested in you towards investigating deeper into this subject.

---

[3] Maxwell's theory of Electromagnetism in 1865 was the first "Unified field theory"

With "The Scientific Proof of God", Swaroop integrates into the standard model of physics the dimensions of Consciousness and Intention. He explains his views on how the universe began and illuminates the twelve principal mysteries presently unanswered in standard, classic physics. Swaroop offers a new application of Einstein's famous equation E=mc2, and how this equation when applied to the universe's dimension of consciousness becomes the Unity Field Law of Causation. *"This one equation explains the cause and effect of any action as well as how primal singularity, the seed of the universe, came into existence before the Big Bang".* As a golden bridge between Science and Spirituality, this book also explains how the name of God can be used for attaining the Unity Consciousness. We are one with the Unity Consciousness, and realizing this, one realizes that it is the goal of life itself.

All ancient languages' alphabets, without exceptions, have attributed to each letter a mysterious corresponding number and the study of that code is called Gematria (hence numerology). Pythagoras said the number is the within of all things.

Genetics, in simple words the study of genes, states that DNA is composed of 4 elements: Hydrogen, Nitrogen, Oxygen and Carbon (H, N, O, C). Carbon is what makes us physical and earthly beings. Now, all this elements have attributed numbers representing their atomic mass, in the Mendeleev table of elements and they equate to the mysterious letters in the ancient alphabets. What that means is that when you look at the human DNA or the DNA of any terrestrial life the numbers that equate to these become 1, 5, 6 and 3 (H 1, N 5, O 6, C 3). This literally reads GOD ETERNAL WITHIN THE BODY! It is in Hebrew, Arabic and Sanskrit three of the root languages.

Plato's, in the ancient Greece said that "the Great God always (practices) geometry" (Αει ο Θεός ο μέγας γεωμετρει - Ai o Theos o Megas geometri) and Pythagoras even postulated "God is a number!" But the most interesting thing about this phrase is that when we count the letters of each word and translate them in numerology, than the meaning of this phrase is Αεί (3,) ο (1) Θεός (4) ο (1) Μέγας (5) γεωμετρεί (9), giving us the exact value of the number π=3,14159... Plato's well-known phrase (Aei o Theos o Megas geometrei), was supplemented in recent times by the Professor of Mathematics N. Hatzidaki (1872-1942) and it became:

Αεί (3,) ο (1) Θεός (4) ο (1) Μέγας (5) γεωμετρεί (9), το (2) κύκλου (6) μήκος (5) ίνα (3) ορίση (5) διαμέτρω (8), παρήγαγεν (9) αριθμόν (7) απέραντον (9), και (3) όν (2), φεύ (3), ουδέποτε (8) όλον (4) θνητοί (6) θα (2) εύρωσι (6) and the following number is obtained: 3.1415926535897932384626.

For those of you more curious or interested to dig deeper for more, here is also a related website for the number π (www.piday.org), in which you will also find a countdown for its feast day.

As Gregg Braden says "If there is an intentionality underlying life itself and human life specifically, it would make tremendous sense to me that whom or whatever is responsible, would somehow have left the sign! Why place that signature into something that can crumble over a few thousand years like a temple wall or into the pages of a book that can be destroyed why not put it into the creation itself!"

"Mapping the chemical sequences for human DNA - the chemical letters that make up the recipe of human life – is a breakthrough that is expected to revolutionize the practice of medicine by paving the way for new drugs and medical therapies" says a web site. The discovery has lasting physical and spiritual implications. A direct link can easily be found between the building blocks of life and the Creator of the universe.

Mankind is fearfully and wonderfully made, with a hidden code within the cell of every life. This code is the alphabet of DNA that spells out the Creator's name and men's purpose. Scientists discovered the map of the four DNA bases that carry the ability to sustain life. These bases, known as chromosomes, are paired differently for each person. Human DNA contains 23 pairs of chromosomes made off of H, N, O, C and their acidic counterparts. Encoded within these elements is an amazing blueprint of life that proves the Creator has put for His own unique stamp upon every person. This stamp is, as mentioned before, the Creator's Name as revealed to Moses thousands of years ago.

As if this is not enough already, things go even deeper with Terrence Howards' lecture at Oxford University: "we know now that energy doesn't die, it reboots itself. So, we are eternal! So, stop panicking thinking your life is over. You've done this trillions of time". It would be interesting to check his theory about numbers and his view upon math and the need for its reform!

The structure of the DNA is a three-dimensional double helix. The two helices are joined together by hydrogen bonds and comprise a sugar-phosphate backbone and the nucleotide bases (guanine, cytosine, adenine and thymine).

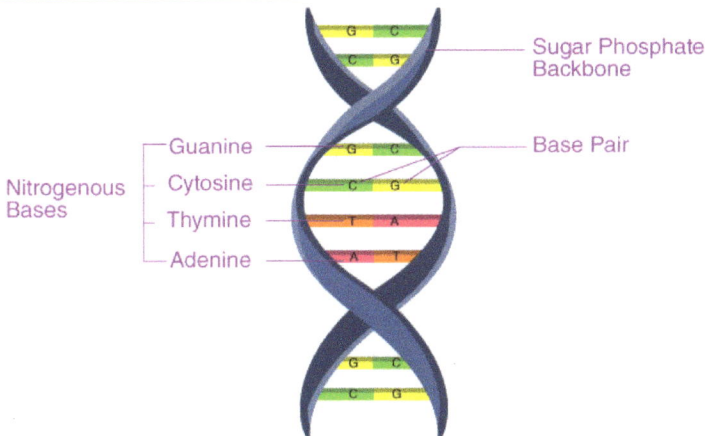

DNA STRUCTURE

BYJU'S
The Learning App

Sugar Phosphate Backbone

Base Pair

Nitrogenous Bases
Guanine
Cytosine
Thymine
Adenine

Scientists used to think that the body transferred or under-stood the energy information through electricity but they found out that it's actually through frequency. And this is what the sci-ence of cymatics states. Cymatics is defined as the study of wave phenomena, especially sound, and their visual representations. Dr. John Stuart Reid, an acoustic-physics scientist points out that since sound underpins almost all matter in the universe and was a potent force in the creation of life in the primordial oceans, it also carries the power to heal life. In his frequent lectures, he reveals ground-breaking information on the mechanisms that underline

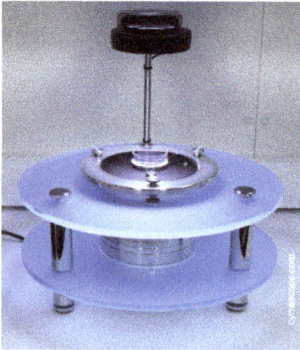

sound therapy, and how it can be applied to improve health naturally. Reid's career in acoustics has spanned five decades and he is widely acknowledged as an authority in cymatics science and speaks on this sub-ject at conferences in Europe and America.

His Cyma Scope invention (2002) has changed our perception of sound forever: seeing sound allows us to understand this omnipresent aspect of our world and universe fuller and deeper. Using this instrument cymatics therapists direct healing frequen-cies into the body to restore resonance and harmony. The healing frequencies are related to those emitted by a healthy organ or body part. In this way, cymatics healers say, the immune system and other natural regulatory functions are stimulated.

The Cyma Scope

"DNA Sonification refers to the use of audio to convey the informa-tion content of a DNA sequence. Audio is created using the rules of gene expression and codons are played as musical notes. DNA can be processed in one of three different ways to read the open reading frame."

You can check for yourself how this would sound like at Mark Temple's YouTube channel as per picture below.

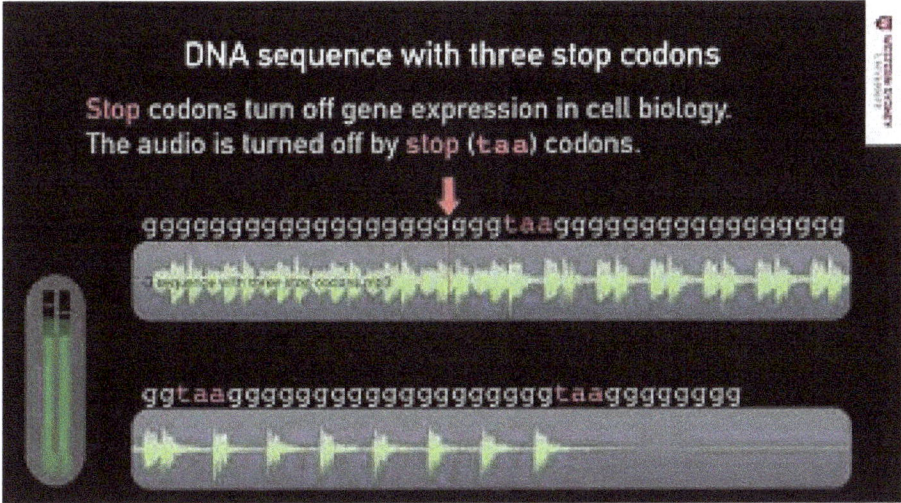

PLAY ALL

So, each and every element has a frequency and sound and when a particular genome tone is played our DNA tightens up and everything else gets pushed out! This means that it's building a *harmonic wave sequencer*! In other words, it's not only that we can reprogram our DNA but also, we can back it up because DNA is actually a hard disk.

George Church and Chris Smith downloaded an e-book into the DNA and then uploaded it to the server and back again. They discovered that we could store data and upload it to and from the DNA and they've managed to replicate the same book 17 billion times into 7 grams of DNA! 1g of DNA stores around 700 Tbites of data! So, basically 13,5 billion years of data, which is approximated as the age of the universe, can be stored in our DNA!

All that was from the beginning is and will still be, nothing was removed or added!

In other words, nothing is new but in a perpetual movement! The only constant in the universe is movement!

In the next subchapters and chapters, I'll take you through the process you can follow to become the better version of yourself, a powerful human being assumed and totally in control of your thoughts, emotions and actions!

CHAPTER 4

——— ·◆◆◆◆◆· ———

Mastering Your Emotions
for Success

Identifying and Overcoming Self-Doubt

Self-doubt is a universal experience that can hinder our growth and success. It is a nagging voice inside of us that questions our abilities, our worth, and our capacity to achieve our goals. In the journey towards becoming a powerful and complete person, it is crucial to address and overcome these doubts that hold us back. This subchapter explores the strategies and mindset shifts required to identify and conquer self-doubt, empowering you to navigate life's challenges with confidence and grace. So, let's begin by:

1. Understanding the Roots of Self-Doubt: Self-doubt often stems from past experiences, societal expectations, and fear of failure. When we allow ourselves to explore the sources of our doubts, we can immediately gain clarity on how they impact our lives, enabling us to challenge and overcome them.

61

2. Cultivating Self-Awareness: Self-awareness, as already addressed before, proved to be a powerful tool for personal growth. So, by recognizing our strengths, weaknesses, and limiting beliefs, we can begin to dismantle self-doubt. Regular self-reflection and journaling can help uncover patterns and triggers that fuel our doubts.

3. Embracing Imperfection: perfectionism and self-doubt often go hand in hand. When we come to realize that is ok to be exactly the way you are, shifting our perspective and embracing imperfection can help us release the pressure to be flawless and instead focus on progress and growth. Celebrating small victories along the way will boost our confidence and motivation. Praise yourself; congratulate yourself for your successes!

4. Surrounding Yourself with Supportive Communities: Building a strong network of like-minded individuals who believe in our abilities is essential. Seek out communities of human beings who uplift and inspire, where you can share your experiences, receive support, and learn from others who have overcome self-doubt. Surround yourself with positive thinkers!

5. Setting Realistic Goals and Taking Action: Break down your goals into manageable steps and take action consistently. Each small achievement will reinforce your belief in yourself and gradually diminish self-doubt. Always celebrate your progress along the way in order to maintain motivation.

6. Practicing Self-Compassion: Self-doubt often arises from a lack of self-compassion. Be gentle with yourself, forgive your mistakes, and treat yourself with kindness. Self-compassion allows us to acknowledge our worthiness and embrace our authentic selves.

7. Seeking Professional Support: If self-doubt feels overwhelming or persists despite your efforts, you should consider

seeking guidance from a specialist like a therapist, a counselor, a coach. They can provide tailored strategies, tools, and support to help you overcome deep-rooted doubts and develop a stronger sense of Self.

Once you implement these strategies and adopt a growth mindset, you will be able to identify and overcome self-doubt, ultimately becoming the empowered person you aspire to be. Remember, you have the power within you to navigate through life's challenges with confidence and grace. Embrace your journey of self-improvement and let go of self-doubt to create a life filled with purpose, fulfillment, and success. You deserve all this!

Embracing Your Unique Voice

In a nowadays world filled with noise and expectations, it can be really challenging to find and express our authentic Self. But as human beings seek to become powerful and complete, it is essential to embrace our unique voice.

Self-improvement through self-awareness is a journey that requires deep introspection. It involves peeling back the layers of all societal conditioning and external influences to uncover your true essence. Embracing your unique voice begins by recognizing your strengths, passions, and values. So, take your time to reflect on what makes you special and what brings you joy. By understanding yourself better, you can create a solid foundation for personal growth.

Embracing your unique voice also means embracing your imperfections. Perfection is an illusion that can hinder personal growth. Recognize that your quirks and flaws are what make you human and relatable. By accepting and embracing them, you allow yourself to be vulnerable and authentic, which can be incredibly empowering.

But owning your unique voice also involves setting boundaries and speaking up for yourself. Too often, we are silenced or overshadowed by societal expectations. Through assertively expressing your thoughts, feelings, and needs, you assert your value and worth. Remember that your voice matters, and your perspective is valid. We all, each and every one of us, hold our own perspective of truth!

As you embrace your unique voice, you become a beacon of inspiration for others. Your authenticity and confidence will inspire those around you to do the same, your story is important and you can make it matter. Dare to share your journey and experiences and you can create a ripple effect of empowerment and change. Be an inspiration.

In conclusion, embracing your unique voice is a powerful act of self-improvement and self-awareness. By understanding and valuing your individuality, you can create a solid foundation for personal growth. Embrace your imperfections, set boundaries, and speak up for yourself. Your unique voice matters, and it has the power to inspire and empower others.

Setting Achievable Goals

In order to become a powerful and complete person, it is essential to set achievable goals that align with your values, desires, and aspirations. Setting goals not only provides direction and purpose in life but also empowers you to take control of your own destiny, thereby transforming your dreams into a reality.

There are a few steps that you can follow:

1. Define Your Vision: Begin by identifying your long-term vision. What do you ultimately want to achieve? Whether it's excelling in your career, maintaining a healthy lifestyle, or nurturing relationships, your vision will serve as a guiding light throughout your journey.
2. Break It Down: Once you have a clear vision, break it down into smaller, manageable goals. All these mini-goals act as stepping-stones towards your larger vision, making it easier to track progress and stay motivated.
3. S.M.A.R.T. Goals: Use the SMART goal-setting framework to ensure your goals are specific, measurable, attainable, relevant, and time-bound.

Specific...

Measurable...

Attainable...

Relevant..

Time-bound..

For example, instead of setting a vague goal like "I want to get fit," set a SMART goal such as "I will exercise for 30 minutes, five days a week, for the next three months to improve my overall fitness level."

4. Prioritize and Focus: It's important to prioritize your goals based on their significance and urgency. Trying to tackle too many goals at once can lead to overwhelm and burnout. Focus on one or two goals at a time and give them your undivided attention.
5. Visualize Success: Visualize yourself already achieving your goals. This powerful technique helps to reinforce your belief in yourself and the possibility of accomplishing what you set out to do. So, sit for a period of time, as long as you consider necessary, in an armchair or lying on the bed in a room,

with relaxing music and start thinking about what you want but from the perspective of the goal already accomplished. See in your mind's eye how you look, how you are dressed, what you feel now that your goal has already been accomplished, what you think and what you decide to do next. By continuing to ask yourself these questions, your mind will have nothing else to do but search for the answers as if you were already in that reality. Then, stand up and write down everything you have visualized, in the smallest details. In the next 21 days, repeat as often as you can, especially in the morning when you wake up and before going to bed, what you have written down, make sure that everything is in the present tense. At first, it may seem that it doesn't make sense, but I promise you that after a period of time, with perseverance, things will change radically so, let yourself be pleasantly surprised by the universe! Create a vision board or write affirmations to remind yourself of your potential and capabilities. You can use methods like the 3-6-9 one: write an affirmation in your journal 3 times in the morning, 6 times at noon and 9 time in the evening, best before going to sleep for 21 days in a row and see what happens!

6. Take Action: Goal setting is meaningless without action. Break down your goals into actionable steps and create a plan to execute them. Consistency and perseverance are the keys to achieving any goal. Celebrate small wins along the way to stay motivated and boost your confidence.

7. Review and Adjust: Regularly review your progress and adjust as necessary. Sometimes circumstances change, and goals need to be adapted. Be flexible and open to modifying your goals while staying true to your overall vision.

Remember, setting achievable goals is not just about accomplishing external milestones; it's also about personal growth, self-awareness, and self-improvement. Embrace the journey, learn from challenges, and celebrate your progress. With confidence and perseverance, you can become the empowered person you aspire to be.

Cultivating a Positive Mindset

In the journey towards empowerment and becoming the best version of yourself, one of the most crucial aspects to focus on is cultivating a positive mindset. The power of positive thoughts and beliefs cannot be underestimated, as they shape our actions, decisions, and ultimately, our lives. In this subchapter, we will explore how human beings can harness the strength of their minds to overcome challenges, enhance self-awareness, and unlock their true potential.

The first step towards cultivating a positive mindset is to become aware of our thoughts and the impact they have on our lives. This self-awareness allows us to identify negative thought patterns in order to replace them with positive ones. Choosing consciously to focus on empowering thoughts, persevering we can build resilience and overcome obstacles.

A key tool in developing a positive mindset is practicing gratitude. When we acknowledge and appreciate the blessings in our lives, we shift our focus away from negativity and embrace a more optimistic outlook. It's like when we take our horses towards a new direction. Regularly expressing gratitude not only enhances our mental wellbeing but also attracts more abundance and positivity into our lives.

Another powerful technique to cultivate a positive mindset is using affirmations. As mentioned before, repeating positive statements about our capabilities, and ourselves we rewire our subconscious mind to believe in our strengths and potential. Affirmations act as a constant reminder of our worthiness and empower us to pursue our dreams fearlessly. Surrounding ourselves with a supportive and uplifting community is also essential in fostering a positive mindset. Connecting with like-minded people who share similar aspirations creates an environment of encouragement, motivation, and inspiration. Through collaboration and shared experiences, we can cultivate a mindset of growth and resilience.

Additionally, it is important to nurture self-care practices that promote mental and emotional wellbeing. Engaging in activities such as meditation, keeping a journal, and exercise helps to reduce stress, boost self-confidence, and enhance overall positivity. Taking time for ourselves allows us to recharge and maintain a healthy mindset amidst life's challenges.

Lastly, embracing a growth mindset is fundamental in our journey towards empowerment. Recognizing that setbacks and failures are opportunities for learning and growth enables us to bounce back stronger and more resilient. By reframing challenges as stepping-stones towards success, we develop a mindset that is open to new possibilities and unafraid of taking risks.

In conclusion, cultivating a positive mindset is an integral part of an empowered human being's journey. By becoming aware of our thoughts, practicing gratitude and affirmations, surrounding ourselves with a supportive community, nurturing self-care practices, and embracing a growth mindset, we can conquer life's challenges and reach our goals. The power of becoming potent and complete lies within each and every one of us, waiting to be unleashed through the cultivation of a positive mindset.

Developing Your Self-Esteem

Self-esteem is another vital component of personal growth and empowerment. It is the foundation upon which we build our confidence and resilience. Self-improvement through self-awareness is a journey that requires us to develop a deep understanding of ourselves and our worth. Many people face societal pressures and expectations that can erode their self-esteem over time. However, by consciously investing in self-care and self-compassion, we can rebuild and strengthen our sense of Self.

The first step towards nurturing your self-esteem is cultivating self-awareness. Take the time to reflect on your strengths, values, and passions. Understand that you are a unique individual with limitless potential. Embrace your imperfections and recognize that they do not define you; instead, they contribute to your authenticity. Once you have gained self-awareness, practice self-compassion. Treat yourself with kindness and empathy, just as you would a dear friend. Acknowledge your accomplishments and learn from your failures without self-judgment. Remember, you are deserving of love and respect, and that starts with Self-love. Surround yourself with a supportive community of like-minded individuals who uplift and celebrate you. Seek out mentors and role models who inspire you to reach your goals. Engage in positive affirmations and visualization exercises to reinforce your belief in your abilities.

As you work on nurturing your self-esteem, remember to set realistic goals and celebrate small victories along the way. Focus on personal growth and progress rather than comparing yourself to others. You are a unique individual and your journey is unique; your success is measured by your own standards, not society's one.

In conclusion, nurturing your self-esteem is a crucial aspect of becoming a powerful and complete person. By cultivating self-awareness, practicing self-compassion, and embracing a supportive community, you can develop unshakable confidence and grace. Remember, you are worthy of all the love, success, and happiness that life has to offer. Empower yourself, and the world will follow suit.

Exploring Emotional Intelligence

Emotional intelligence, often referred to as EQ, is defined as the ability to recognize, understand, and manage our own emotions, as well as effectively navigate and empathize with the emotions of others. It is a skill that plays a pivotal role in our relationships, decision-making, and overall wellbeing. As human beings seeking empowerment, developing emotional intelligence is essential. We are empathic beings and self-awareness serves as the foundation of emotional intelligence. It involves becoming attuned to our own emotions, recognizing patterns, triggers, and habitual responses. After gaining this deep understanding of ourselves, we can then begin to manage our emotions effectively.

According to Daniel Goleman, the psychologist who popularized the term 'emotional intelligence', EQ consists of:

- Self-awareness.
- Self-regulation.
- Motivation.
- Empathy.
- Social skills.

Furthermore, exploring emotional intelligence entails developing our empathy and understanding towards others. As human beings seeking to empower ourselves, we should foster meaningful connections and build strong relationships. The proper guidance on how to cultivate empathy, active listening, and effective communication skills enables human beings to create harmonious connections both personally and professionally. Additionally, emotional intelligence plays a significant role in decision-making. By understanding and managing our emotions, we can make informed choices that align with our values and goals. Ultimately, the exploration of emotional intelligence empowers human beings to embrace their authentic selves, navigate relationships with wisdom and grace, and make sound decisions that lead to personal and professional fulfillment. By integrating emotional intelligence into their lives, we can unlock our true potential and become powerful, complete individuals.

As you embark on this journey of self-improvement and empowerment, this subchapter will serve as a guide, offering valuable insights and practical tools to enhance your emotional intelligence. Remember, emotional intelligence is not just an innate trait but also a skill that can be developed and honed. Through practice and perseverance, you have the strength to unlock the empowering benefits of emotional intelligence and become the individual you are aiming to be.

Empathy serves as a powerful tool for self-improvement through self-awareness. By putting ourselves in others' shoes, we can gain a deeper understanding of their experiences, emotions, and perspectives. This practice cultivates a heightened sense of empathy towards us as well. When we acknowledge and validate our own feelings and struggles, we embark on a journey of self-compassion, which is essential for personal growth.

Compassion, on the other hand, allows us to extend empathy beyond understanding and into action. As empowered human beings, we have the opportunity to make a positive impact on the lives of others and our communities. By showing compassion to those in need, we create a ripple effect of kindness and support. In doing so, we not only uplift others but also enhance our own sense of purpose and fulfillment.

Furthermore, enhancing empathy and compassion enables us to build stronger and more fulfilling relationships. As we develop a genuine interest in others' wellbeing, we foster deeper connections based on trust, understanding, and support. These relationships become a source of strength and encouragement, propelling us towards our goals and helping us overcome life's challenges with grace.

There are some practical strategies for enhancing empathy and compassion in our daily lives. People who were not empathic are often seen as cold and self-absorbed and they lead isolated lives. Sociopaths are lacking in empathy. Conversely, people who are empathetic are perceived as warm and caring. Studies have shown that empathy is partly innate and partly learned. We will further explore few techniques for improving oneself and strengthen your empathy like the ones suggested by coach Andrew Sobel[4]:

1. **Challenge yourself.** Undertake challenging experiences, which push you outside your comfort zone. Learn a new skill, for example, such as a musical instrument, hobby, or foreign language. Develop a new professional competency. Doing things like this will humble you, and humility is a key enabler of empathy.

2. **Get out of your usual environment.** Travel, especially to new places and cultures. It gives you a better appreciation for others.

3. **Get feedback.** Ask for feedback about your relationship skills (e.g., listening) from family, friends, and colleagues— and then check in with them periodically to see how you're doing.

4. **Explore the heart not just the head.** Read literature that explores personal relationships and emotions. This has been shown to improve the empathy of young doctors.

5. **Walk in others' shoes.** Talk to others about what it is like to walk in their shoes—about their issues and concerns and how they perceived experiences you both shared.

---

[4] Andrew Sobel, is a well renowned author, an executive educator and coach as well as strategy advisor to senior management

6. **Examine your biases.** We all have hidden (and sometimes not-so-hidden) biases that interfere with our ability to listen and empathize. These are often centered around visible factors such as age, race, and gender. Don't think you have any biases? Think again—we all do.

7. **Cultivate your sense of curiosity.** What can you learn from a very young colleague who is "inexperienced?" What can you learn from a client you view as "narrow"? Curious people ask lots of questions (point 8), leading them to develop a stronger understanding of the people around them.

8. **Ask better questions.** Bring three or four thoughtful, even provocative questions to every conversation you have with clients or colleagues.

If you want to develop your skill towards more area of your life it is of outmost importance to learn to empathize and build the relationships that truly matter to attain what you desire, from personal development to career success!

By embracing empathy and compassion, we not only transform our own lives but also contribute to the betterment of society as a whole. As powerful and complete human beings, we have the ability to create positive change by nurturing these qualities within ourselves. Together, empowering others and ourselves with the beauty of empathy and compassion we can build a better and more fulfilled life and world!

Managing Stress and Anxiety

In our fast-paced and demanding world, stress and anxiety have become all too familiar. As human beings striving to become powerful and complete, it is essential to equip ourselves with effective strategies to manage these overwhelming emotions. Beyond the health consequences these two emotions are known as the most destructive ones.

Stress is perceived as the illness of our century and it is blamed for most of our sufferance. When the hormones of stress are released in our body, all our systems are disturbed. Our body is made to react to stress in ways meant to protect us against threats from predators and other aggressors. Such threats are rare today. It is the so-called "reptilian brain" that is signaled and acts as such. But that doesn't mean that life is free of stress. An example of a perceived threat is a large dog barking at you during your morning walk. Through nerves and hormonal signals, this system prompts the adrenal glands, found on top of the kidneys, to release the two hormones: adrenaline and cortisol. The body's stress response system is usually self-limiting. But once a perceived threat has passed, hormones return to typical levels. And while adrenaline and cortisol levels drop, your heart rate and blood pressure return to typical levels, all other systems go back to their regular activities.

Adrenaline makes the heartbeat faster, causing blood pressure to go up and giving you more energy. Cortisol, the primary stress hormone, increases sugar, also called glucose, in the bloodstream, enhances the brain's use of glucose and increases the availability of substances in the body that repair tissues.

Cortisol also slows functions that would be nonessential or harmful in a fight-or-flight situation. It changes immune system responses and suppresses the digestive system, the reproductive system and growth processes. This complex natural alarm system also communicates with the brain regions that control mood, motivation and fear. This is the body's reaction to threat, it turns on the defense mechanisms setting the rest of the systems on the survival mode, pumping the blood towards the members to activate the fight-or-flight! But when stressors are always present and you always feel under attack, that fight-or-flight reaction remains turned on.

Now, the long-term activation of the stress response system and too much exposure to cortisol and other stress hormones can disrupt almost all the body's processes. This puts you at higher risk of many health problems, including anxiety, depression, digestive problems, headaches, muscle tension and pain, heart disease, heart attack, high blood pressure and stroke, sleep problems, weight gain, problems with memory and focus.

That's why it's so important to learn healthy ways to cope with your life stressors.

The way we react to a potentially stressful event is different from person to person. How you react to your life stressors is affected by different factors as:

- Genetics. The genes that control the stress response keep most people at a fairly steady emotional level, only sometimes priming the body for fight or flight. More active or less active stress responses may stem from slight differences in these genes.
- Life experiences. Strong stress reactions sometimes can be traced to traumatic events. People who were neglected or abused as children tend to be especially at risk of experiencing high stress. The same is true of airplane crash survivors, people in the military, police officers and firefighters, and people who have experienced violent crime.

But there are people that seem relaxed about almost everything. You may have friends who react strongly to the slightest stress or others that are calm and relaxed. Most people react to life stressors somewhere between those extremes.

Stressful events are life facts. And no matter what your reaction is, you may be able to change your current situation not only about yourself but also about others by offering them a healthy way of helping. And you can learn what steps to take to manage the impact these events have on you. In the end, stress is a natural mechanism we are equipped with, as a defense response to a possible life threat! The exposure to stress, in terms of duration, makes it dangerous and destructive.

First of all, you can learn to identify what causes you stress. And then you can learn how to take care of yourself physically and emotionally in the face of stressful situations.

Suggestion for stress management:

- Eat a healthy diet and get regular exercise. Get plenty of sleep but quality sleep too.
- Do relaxation exercises such as yoga, deep breathing, massage or meditation.
- Keep a journal. Write about your thoughts or what you're grateful for in your life.
- Take time for hobbies, such as reading or listening to music. Or watch your favorite show or movie.
- Foster healthy friendships and talk with friends and family.
- Have a sense of humor. Find ways to include humor and laughter in your life, such as watching funny movies or looking at joke websites.
- Volunteer in your community.
- Organize and focus on what you need to get done at home and work and remove tasks that aren't needed.
- Seek professional counseling. A counselor can always help you learn specific coping skills to manage stress.

Stay away from unhealthy ways of managing your stress, such as using alcohol, tobacco, drugs or excess food. And always ask for professional help if you feel you are overwhelmed.

There are many rewards for learning to manage stress. For example, you can have peace of mind, fewer stressors and less anxiety, a better quality of life, improvement of health conditions such as blood pressure, better self-control and focus, and better relationships. And it might even lead to a longer, healthier life.

Vitality Tone and Attitude Scale

Increases in strength and energy reserves	Emotion · Vibration	Thoughts · Motivation	Attitude Likely Held
Recharge and Refresh · Muscles firm Energy gains from Even to Positive. Region of Emotions characterized as Pleasurable or Good. Blood flows with vigor.	Serenity	Peace · Aliveness · Well Being	"Wow, this is fascinating. I'm alive and learning."
	Joy · Enthusiasm	Welcome · Exhilaration · Abundance	
	Compassion	Empathy · Inspiration · Clarity	
	Appreciation	Gratitude · Devotion · Generosity	
	Love	Cooperation · Trust	
Balance - Even - Stable - Relief	Satisfaction	Amusement · Curiosity	"Thank goodness, here this comes up for my healing."
	Power · Strength	Discovery · Challenge	
	Self Esteem · Dignity	Duty · Obligation	
	Neutral · Acceptance	Contentment-Safety-Aplomb	
	Glee · Happy	Nervous · Worry · Hyper	
	Surprise · Shock	Confusion · Annoyance	
Muscle Release - Energy moves from Negative to Even. Region of emotions characterized those called Painful or Sad. Moving downward	Anger	Rage · Defiance · Boredom	"Oh no, not again."
	Guilt	Resentment · Remorse	
	Fear	Threat · Hate · Blame	
	Sadness · Grief	Depleted · Loss · Burden	
	Hopelessness	Resignation · Depressed	
	Numb · Powerlessness	Overwhelm · Frozen	
	Shame	Apathy · Helpless · Death	
Muscles tighten constrict or atrophy.			2006, 2010, 2013© Stephen J. Cocconi. Graphic design: Ardis Bow.

When dealing with anxiety, we should focus upon a few aspects like:

1. Understanding Stress and Anxiety: Begin by recognizing the difference between stress and anxiety. Stress is a natural response to external pressures, while anxiety is an internal feeling of unease or worry. By understanding the root causes, you can identify triggers and develop appropriate coping mechanisms.

2. Self-Awareness: Developing self-awareness is crucial in managing anxiety but also stress. Take time to reflect on your emotions, thoughts, and physical sensations. This awareness will help you identify early signs of anxiety, allowing you to intervene before they escalate.

3. Mindfulness and Meditation: Incorporating mindfulness and meditation practices into your routine can be immensely beneficial. These techniques help anchor you in the present moment, reducing anxiety and promoting a sense of calm. Listening to the sound of the ocean waves or rain can help relax.

4. Healthy Lifestyle: Nurturing your physical wellbeing is vital for managing both stress and anxiety. So, engage in regular exercise, eat a balanced diet, and prioritize restful sleep. Taking care of your body enhances your ability to cope with anxiety triggers effectively.

5. Positive Self-Talk: Transform negative self-talk into positive affirmations. Remind yourself of your strengths, accomplishments, and capabilities. This shift in mindset will empower you to face challenges with confidence, minimizing anxiety.

6. Time Management: Feeling overwhelmed often contributes to anxiety. Learn to prioritize tasks, delegate when possible,

and set realistic goals. By managing your time effectively, you can reduce the pressure and regain a sense of control.

7. Seeking Support: Remember, you don't have to face stress and anxiety alone. Reach out to trusted friends, family, or professionals who can provide guidance and support. Joining a community of like-minded people, or self-improvement groups, as well as therapy, coaching can be immensely helpful.

By implementing these strategies, you will enhance your ability to manage stress and anxiety effectively. Embrace self-improvement through self-awareness, and let it be your guiding light on this transformative journey. Remember, you have the strength within you to overcome any obstacle and create a life filled with joy, purpose, and fulfillment.

Building Resilience in the Face of Challenges

Introduction:

In the journey towards empowerment and self-improvement, challenges are inevitable. Life's hurdles can often leave us feeling overwhelmed and uncertain of our capabilities. However, building resilience can help us navigate through these challenges with confidence and grace.

Understanding Resilience:

Resilience is perceived as being an ability to bounce back from adversity, to adapt and grow stronger through life's challenges. It is a quality that enables us to face setbacks, overcome obstacles, and emerge stronger on the other side. Resilience is not about avoiding difficulties but rather about developing the skills and mindset to deal with them effectively and become stronger:

1. Cultivating Self-Awareness: The first step towards building resilience is developing self-awareness. By understanding our strengths, weaknesses, and triggers, we can better prepare ourselves for the challenges that lie ahead. Through introspection and reflection, we can identify our coping mechanisms, values, and goals, enabling us to make conscious choices that align with our true selves.

2. Embracing a Growth Mindset: Resilient individuals believe in the power of growth and learning. They view challenges as opportunities for personal development and approach setbacks as stepping-stones towards success. By adopting a growth mindset, we can reframe our perspectives, seeing failures as temporary setbacks rather than defining moments, and nurturing a sense of optimism and possibility.

3. Building a Supportive Network: A strong support system is essential when facing challenges. Surrounding ourselves with like-minded individuals who uplift and inspire us can provide the encouragement and strength we need during difficult times. Building connections through mentorship, coaching, or joining communities of empowered human beings can offer guidance, inspiration, and a shared sense of purpose.

4. Practicing Self-Care: Resilience is not solely about overcoming challenges; it also involves taking care of ourselves. Prioritizing self-care helps us maintain emotional and physical wellbeing, enabling us to face challenges with a clear mind and an open heart. Engaging in activities such as mindfulness, exercise, journaling, and self-reflection can nurture our resilience and provide us with the energy needed to navigate life's ups and downs.

As a Conclusion: Building resilience in the face of challenges is a lifelong journey that requires commitment, self-awareness, and a growth mindset. By cultivating these qualities and implementing practical strategies, human beings seeking empowerment and self-improvement can navigate life's challenges with confidence and attain wisdom. Remember, resilience is not about avoiding difficulties, but about embracing them as opportunities for growth and personal transformation. If we consider resilience as our foundation, we can become powerful and complete individuals, ready to conquer anything that comes our way.

CHAPTER 5

———— ·◆◆◆◆·· ————

Creating a Life Vision: Designing Your Path to Fulfillment

Defining Your Values and Priorities

In the journey towards empowerment and personal growth, it is crucial to define your values and priorities. Understanding what truly matters to you and aligning your actions and decisions accordingly can empower you to live a more fulfilled and purpose-driven life. As a woman who's constantly seeking to become the best version of myself, strong and fulfilled, I will guide you through this process by establishing a strong foundation for personal transformation.

First and foremost, it is essential to explore the concept of values. Values, as mentioned in the first chapter, are the guiding principles that shape our beliefs, attitudes, and behaviors. They represent what we hold most dear and what we consider most important in our life. By identifying your core values, you gain clarity about what truly matters to you, enabling you to make choices that are in alignment with your authentic self. So, when you identify those values that represent you the most you get a clearer picture of the person you are values like family, truth, love, wellbeing, wealth or abundance, joy, happiness and so on can be at the top of your list.

Next, we delve into the process of determining your priorities. Life is abundant with opportunities and demands, and it is easy to get overwhelmed or lose focus. *Where attention goes, energy flows!* So, understanding and setting your priorities leads you to take meaningful decisions that bring you closer to your goals and aspirations. Identifying your priorities helps you allocate your time, energy, and resources more effectively, ensuring that you are investing in what truly matters to you. By analyzing and exploring your values and priorities, you will gain insights into your passions, strengths, and areas for growth. This self-reflection will serve as a solid foundation for designing a life that aligns with your values and empowers you to create the future you desire.

But of course, this journey doesn't have to be done all alone and asking for guidance and help will only improve the results and eventually diminish the period till completion. Therefore, we will explore a bit the concept of professional coaching and its potential to support your personal growth journey. Life coaching provides a structured and supportive environment for self-improvement, helping you uncover your inner strengths, overcome obstacles, and achieve your goals. By embracing the power of coaching, you can accelerate your progress and unlock your full potential.

As you strive for empowerment, it is vital to prioritize your own physical, emotional, and mental health. By nurturing yourself, you can show up as the best version of yourself, ready to tackle life's challenges.

In conclusion, defining your values and priorities is the cornerstone of personal empowerment. By gaining clarity on what truly matters to you, setting meaningful goals, and prioritizing self-care, you can navigate life's challenges with confidence and grace. This subchapter will serve as your guide, supporting you on your journey of self-improvement and self-awareness. Remember, you have the power to become the empowered being you aspire to be.

Setting Meaningful Life Goals

In the journey towards empowerment and self-fulfillment, one of the most crucial steps is setting meaningful life goals. I cannot underline how important it is to have a clear vision and purpose in our lives. It is essential to set goals that align with our values and desires, in order to help us navigate our life's challenges easier and achieve our scope. So, how to go through the process of setting meaningful life goals, and take charge of your destiny:

The first step in setting meaningful life goals is to understand who you are through self-introspection and awareness: reflect on your passions, values, and aspirations; take time to get a clear picture of "What brings you joy?" identify "What are your core beliefs?" Understanding yourself on a deeper level will enable you to set goals that are truly meaningful to you. Consider your strengths and weaknesses, as well as the areas of your life where you would like to see growth and improvement. Write them down, on a piece of paper, in two separate columns, one next to the other and compare them.

Once you have gained clarity and made the click, it is time to define your goals. Remember, meaningful goals are those that resonate with your values and bring a sense of purpose to your life. Whether it is pursuing a fulfilling career, improving your relationships, or enhancing your wellbeing, your goals should reflect what matters most to you.

To ensure your goals are achievable and realistic, break them down into smaller, manageable steps and easier to attain. This will prevent overwhelm and will allow you to track your progress along the way. Set specific deadlines and create a timeline for each step, holding yourself accountable to stay on track.

For ex. you want to lose weight: write down the main goal like in it is already fulfilled:

I am... kg until the end of this year and I stay fit and healthy!

Then break it down to smaller steps:

By the end of this weak I loose in a healthy and safe way .... kg and every day I'm feeling better and better.

Please be aware not to exaggerate, set attainable goals, so that you get to accomplish them and every time remember to celebrate each victory, no matter how small! You can ask a friend or relative to partner with you and hold you accountable at least at the begging of your journey towards your goal.

Additionally, it is important to regularly review and reassess your goals. Life is ever evolving, and your aspirations may change over time. By periodically evaluating your goals, you can ensure they remain relevant and aligned with your evolving self.

Lastly, surround yourself with a support system that encourages and uplifts you. Seek out mentors, join communities of like-minded individuals, or consider working with a coach you resonate. Having a strong support network will provide you with guidance, motivation, and accountability as you work towards your goals.

Remember, setting meaningful life goals is a powerful tool that will empower you to live a life of purpose and fulfillment. By aligning your goals with your values and desires, you can easier navigate life's challenges. Embrace this journey of self-improvement and self-awareness, and watch as you transform into the empowered person you were meant to be.

Creating a Vision Board for Clarity

In the journey of becoming a powerful and complete being, one of the most effective tools you can use is a vision board. A vision board is a visual representation of your goals, dreams, and aspirations. It serves as a powerful reminder of what you want to achieve and helps you stay focused, motivated, and aligned with your true desires.

The process of creating a vision board begins with your self-awareness. Take some time to reflect on your life, your dreams, and what truly makes you happy. What areas of your life do you want to improve? What goals do you want to achieve? By gaining clarity on these aspects, you can start creating a vision board that aligns with your deepest desires.

Start by gathering some materials, such as a large poster board, magazines, scissors, glue, and markers. Select images, words, and phrases that resonate with your goals and aspirations. Be open-minded and choose items that speak to your heart, even if they seem unrelated or far-fetched at the moment. Remember, this is your personal vision board, and it should reflect your unique journey.

Arrange the images and words on the poster board in a way that feels right to you. You can organize them by category or create a collage that captures the essence of your dreams. As you glue each item in place, visualize yourself already living the life you desire. Feel the emotions and sensations associated with achieving your goals.

Once your vision board is complete, place it somewhere you can see it every day. This constant visual reminder will keep you focused and motivated to act towards your dreams. Take a few moments each day to look at your vision board and visualize yourself living the life you've created on the board for real. Use this time to reaffirm your commitment and belief in your abilities.

Creating a vision board is just the beginning. This is a powerful tool that can help you manifest your desires, but it requires consistent action and dedication. Use your vision board as a roadmap to guide your decisions and choices. As mentioned before, break down your goals into smaller, actionable steps, and take consistent action towards them. Celebrate each milestone and adjust your vision board as your dreams evolve and manifest.

Remember, you have the power to create the life you desire. With a clear vision and a vision board as your guide, you can definitely reach your desired goal. Embrace this journey and watch yourself transform day by day, little by little, into the person you want to be and were always meant to be.

Taking Inspired Action Steps

In the journey towards discovering your self-awareness, empowerment and personal fulfillment, it is crucial to take inspired action steps. These actions propel us forward, helping us overcome obstacles, and create the life we truly desire. The key lies in understanding that we have the power to make positive changes, and it all starts with self-improvement through self-awareness. This involves paying attention to our thoughts, emotions, and behaviors, and understanding how they shape our reality. By gaining insight into our strengths, weaknesses, passions, and values, we can make informed decisions and take action aligned with our true selves. Once we have developed this self-awareness, the next step is to set clear intentions and goals.

What do you truly want to achieve? What is your vision for your life? Setting specific, measurable, achievable, relevant, and time-bound (SMART) goals will provide you with a roadmap towards success.

Vision Board Planner

MY GRAND VISION:	I CAN BRING THIS VISION INTO FRUITION BY:
	THIS VISION FOR MY LIFE MATTERS BECAUSE:

	WHAT TO DO:	WHY IT MATTERS:
SPIRITUAL GOALS		
CAREER GOALS	WHAT TO DO:	WHY IT MATTERS:
FAMILY GOALS	WHAT TO DO:	WHY IT MATTERS:
PERSONAL GROWTH GOALS	WHAT TO DO:	WHY IT MATTERS:

Whether it's advancing in your career, improving relationships, or enhancing your overall wellbeing, having well-defined goals will bring focus and clarity to your actions.

However, just setting goals alone is not enough. The real magic lies in taking inspired action steps towards these goals. Inspired action is that action taken from a place of inner alignment, intuition, and passion. It involves stepping out of your comfort zone, pushing through fears, and embracing growth. You know what it is said: living it or facing it overcomes fear! Taking action that is in alignment with your authentic Self, will lead you to tap into your personal power and unlock your full potential.

But remember, it is essential to have a support system in place as you embark on this journey. So, seek out mentors, coaches, or like-minded individuals who can provide guidance, encouragement, and accountability. Surrounding yourself with positive influences will keep you motivated and inspired, especially during challenging times.

Taking inspired action steps is not always easy, but it is the key to becoming an empowered person. Believe in yourself, trust your intuition, and have faith in your ability to create the life you desire. Embrace the challenges as opportunities for growth, and never underestimate the power of your actions.

Celebrating Milestones and Progress

In the journey towards empowerment, it is essential to acknowledge and celebrate the milestones and progress we make along the way. Each step forward, no matter how small, brings us closer to becoming the powerful and complete human beings we aspire to be. This subchapter is dedicated to honoring the achievements and recognizing the growth we experience on this transformative path.

As human beings seeking self-improvement through self-awareness, we often encounter challenges that test our resilience and determination. It is crucial to remember that progress is not always linear, and setbacks are a natural part of any personal growth journey. But by celebrating milestones, we embrace the power of positivity and self-appreciation, fueling our motivation to continue striving for greatness.

One way to honor these milestones is through reflection. So, take a moment to look back at where you started to see yourself at that moment in time and recognize the distance you have traveled from that point in time to the present one. Perhaps you embarked on a new career path, started a business, lost weight, or found the courage to break free from toxic relationships. These achievements deserve to be celebrated, as they symbolize your strength and determination to create a life that aligns with your values and aspirations.

Furthermore, celebrating milestones allows us to cultivate a sense of gratitude. It reminds us to appreciate the small victories that often go unnoticed. Whether it is completing a challenging task, overcoming a fear, or simply taking time for self-care, every step taken towards personal growth is worthy of acknowledgment. By expressing our gratitude for these accomplishments, we actually invite more positive energy into our lives, attracting further progress and fulfillment. Remember, the law of attraction works by like attracts like! So, the more grateful and genuine we feel and express the more of the same we will attract in our lives!

Additionally, celebrating milestones can be a powerful source of inspiration for others. As human beings striving for a better life and empowerment, we have the opportunity to uplift and support our fellow humans on their own journeys. By openly sharing our achievements, we create a ripple effect of encouragement and motivation, fostering a community of strong, confident human beings who uplift one another.

In conclusion, celebrating milestones and progress is an essential practice for all those seeking empowerment and self-improvement. It allows us to recognize our growth, cultivate gratitude, and inspire others to embark on their own transformative paths.

So, let us embrace and celebrate every milestone, big or small, as we navigate life's challenges with confidence and grace, empowering ourselves and those around us.

CHAPTER 6

——— ·◆◆ ◆ ◆◆· ———

Cultivating Your Awareness – The Pineal Gland

Understanding the Importance of the pineal gland

In the journey towards empowerment and personal growth, it is crucial to recognize and understand the functions of our bodies with all its the systems, glands and organs. As Homo Sapiens, we are spirit or essence, having a physical experience in a body. So, it is utterly important to understand how the body functions and operates, in order to be able to care for it accordingly and further develop. While seeking to become powerful and complete, we must understand our body's biology for this plays a pivotal role in shaping our lives and entire existence.

But, of course, there is always the choice: we can be ignorant and go about our life without caring about who and what we really are or we can inquire and, if there is even the slightest bit of curiosity, we'll be able to find that we are destined for greater things than just the mere ones we are programmed or thought to do. For example: did you know that we are not supposed to ever get sick? Many hypnotherapies and regressions prove that our illnesses are actually manifested by us, for the mere reason of understanding a certain aspect we are not quite aware of, in that particular moment and are confronted with it! Cancer, as has been scientifically found, appears to be caused by unmanifested anger. So, isn't it normal that once we found the cause to eliminate it? Therefore, we must continuously work with ourselves, be always self-aware and find the appropriate ways and techniques to stay in a positive mind-set.

The pineal gland, also known as the third eye, is believed to help you determine who you are based on your presence in the universe. But a lot of us are coming and leaving this planet without ever achieving that awareness because we are so concerned about what is going on around and outside of us and so much connected to the rules and the structures of society. We are inherently social creatures. But the third eye allows you to be able to form an opinion of you outside of the restrictions and structures of society. So, what it is said and believed is that when your third eye is open you know who you truly are within the context of the universe not within the context of society.

Our relationships with family, friends, romantic partners, and even colleagues greatly influence our emotional wellbeing and overall happiness. When we surround ourselves with positive and supportive individuals, we create an environment conducive to personal growth. Subsequently, these relationships give us the love, encouragement, and guidance we need to overcome challenges and reach our full potential.

However, it is equally essential to be mindful of toxic relationships that drain our energy and hinder our progress. Recognizing the signs of unhealthy dynamics allows us to make informed decisions about whether to invest our time and emotional resources in such connections. But through self-awareness, we can identify which are the patterns that stop us, set boundaries, and choose to prioritize the relationships that uplift and inspire us.

Furthermore, relationships offer us invaluable opportunities for self-reflection and personal growth. By engaging in meaningful conversations and sharing experiences with others, we gain new perspectives and insights that expand our understanding of the world around us and most importantly of ourselves. These inter-actions enable us to challenge our beliefs, confront our fears, and embark on a journey of self-improvement.

But many times, our paradigms keep us prisoners in its' own meanders without us being able to realize what is really happening in our life and what is the cause of our sufferance! And we end up tormenting and struggling thinking of either "what did I do wrong to deserve this?" or "why is this happening to me, why me?" or not even wondering at all and just accepting and confronting those sit-uations had down! But if we take a moment to stop thinking over and over the same thoughts and allow ourselves to breathe and calm our mind, in that moment we actually determine a zero or start-ing point of new possible future. When we allow and perpetuate the same thoughts, we actually are like a dog turning around and chasing his tail but expecting a new perspective. Well, the law of attractions tells us clearly that "like attracts like" so how can some-thing new come out from something old? We must first change our thoughts and then our perspective will change as a consequence! But how can we do that, because it is easy to say it but we must act.

Through specific coaching sessions, we can gain clarity on our values, explore our strengths, and develop effective communica-tion and conflict resolution skills. This self-awareness and personal development journey within enhance not only our own lives but also our relationships with others and, as well as with ourselves.

Affirmations play an important role along with the vision-boards and the visualization techniques and used together and combined they really lead to amazing results. But affirmations alone as well are proven very useful. However, using only "I am" affirmations (I am healthy and strong, I am rich and abundant and so on) and sticking solely to these affirmations, you're operating at the lowest energy level - the root chakra. No wonder it takes you months to attract anything! Each chakra system has its own affirmations; let's take as an example the affirmation of *I am abundant* and go through the chakras with different specific forms for each: start with

1. The Root Chakra: *I am wealthy, I am abundant,* move up to
2. The Sacral Chakra: *I feel well, I feel abundance,* progress to
3. The Solar Plexus Chakra: *I do not chase, I attract,* move onto
4. The Heart Chakra: *I love wealth, I love abundance,* advance to
5. The Throat Chakra: *I speak wealth into existence, continue to*
6. The Third Eye Chakra *I see wealth and abundance all around me,* finally
7. The Crown Chakra *I understand that wealth is easy to attract*

Try practicing these affirmations every day, for seven days straight. Now while all chakras are important, the third eye is for our subject the one I'll talk about explaining why it is so important.

Well, it is believed to be a potent source of intuitive wisdom which can guide you towards creative pursuits, get rid of and stay away from negativity, provide knowledgeable inside and also possibly leading you to the highest form of intelligence, helping open your eyes towards what needs attention. Now, the best method to stimulate and open your third eye is through meditation but also gazing at the sun actually stimulates the pineal gland. But I cannot think of anyone better than Dr. Joe Dispenza as the promoter and actually a researcher of all the process of stimulating and opening the third eye. He conducts specific *third eye opening* meditations and holds courses all around the world, of thousands of dollars for this.

The third eye meditation, just like any form of meditation, requires you to stay put in a calm environment and enjoy the benefits of soothing sounds like Solfeggio's frequencies or alpha-theta sounds. To begin with, start by sitting in a comfortable position on a chair or bed or even on the floor. Keep your spine straight but erect, shoulders relaxed and the hands on the knees. Your stomach, jaws and face should be totally relaxed and open to positive energy.

Start by lightly bringing together the index finger to the thumb and close your eyes, gently. Next, breathe slowly: inhale and exhale through the nose, gently. With your eyes still closed, try to look up at the *third eye* located just between your eyebrows; you can also make use of your fingers to locate the exact point by tapping it slightly. Put your intention to get your pineal gland or third eye, fully stimulated. You may try and do it yourself until succeeding or you may follow dr. Dispenza's *pineal gland guided meditation-third eye activation,* or you can join a group of like-minded and do it together. The power of the group, as dr. Dispenza says as well, is demonstrated to have a great impact subsequently leading to success quicker than being all by yourself.

But let's see and get to understand what this gland's function is in the body. Even from antiquity and also in the antediluvian texts, the ancient Gods were all depicted with conelike structures in their hands. The Sumerian kings and gods are holding in their hands these cones. If we look into a transversal section of the had we'll see that the two glands: the pineal or epiphysis and the pituitary along with the hypothalamus resemble an eye - The ancient Egyptian's *Eye of Horus* or *the Eye of RA -.*

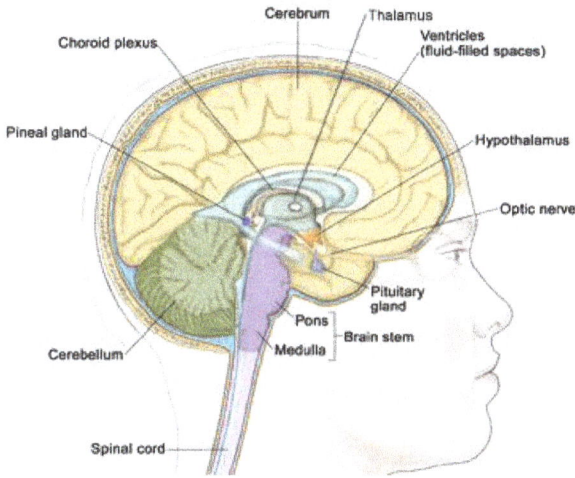

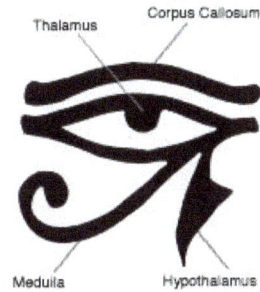

Similarities between the cross section of the brain showing the Pineal Gland and the 'Eye of Ra".

The Eye of Ra, or the Eye of Horus, is regarded today as an Egyptian symbol of protection, royal power and good health.

Horus was the ancient Egyptian sky god who was usually depicted as a falcon, and symbolized enlightenment and awakening (the archetype of the awakened Christ Consciousness) His right eye was associated with the Sun God 'RA' who ruled in all parts of the created world: The Sky, The Earth and The Underworld.

Wikipedia writes about the pineal gland that it "is a small endocrine gland in the brain of most vertebrates. In the darkness the pineal gland produces melatonin, a serotonin-derived hormone, which modulates sleep patterns following the diurnal cycles. The shape of the gland resembles a pinecone, which gives it its name.

The pineal gland is located in the epithalamus, near the center of the brain, between the two hemispheres, tucked in a groove where the two halves of the thalamus join. It is one of the neuroendocrine secretory circumventricular organs in which capillaries are mostly permeable to solutes in the blood. Ancient Greeks were the first to notice the pineal gland and believed it to be a valve, a guardian for the flow of pneuma."

So, what is important for us to remember is that the pineal gland is responsible of the secretion of a hormone called melatonin, which helps our body control day-night sleep patterns and your internal body clock (*circadian rhythms*), as well as the release of reproductive hormones along with temperature control.

Prof. Ashok Sahai and Raj Kumari Sahai, many years before, in the Journal of the Anatomical Society of India an article[5] based upon the research and discoveries about the *Pineal gland: A structural and functional enigma.* They mentioned even from the beginning of the article that: "*The structures and functions of neuroendocrine pineal gland remains an enigma to both philosophers and scientists alike since time immemorial. Recently a neuronal circuit consisting of seven neurons between retina and pineal gland has been established to relate the effect of light and other rays on its secretion. The various physical properties such as piezoelectricity, piezo luminescence, electromagnetic field, solar flare, infrared energy are also explained and correlated with the structural and secretion components of the gland".* And they continue going deeper and more scientifically: "*The presence of all enzymes needed for the synthesis of di-methyl-tryptamine (DMT) in pineal gland explains the near-death experience (NDE) phenomenon. The various audiovisual hallucinations in NDE phenomenon occur due to massive increase of DMT in pineal gland before death. A very high concentration of*

---

[5] The link to the article https://asiindia.in/Previous-Issues/2013%20dec%20issue/Pineal-gland--A-structural-and-functio_2013_Journal-of-the-Anatomical-Societ.pdf

di-methyl-tryptamine (DMT), presence of retinal proteins in 5-10% of pinealocytes, its role in thermoregulation and a possible role as magnetoreceptor in blind men and highest deposits of fluoride in the body are not only interesting but significant for the future research. Hence a lot of further research on pineal gland is still required to correlate its unique properties with its structural components."

Nowadays, dr. Dispenza, through his work and experience with people has demonstrated that the illumination/opening of this "third eye" by decalcification of the gland has tremendous effects and benefits overall for human beings. Second harmonic generation (SHG) measurements showed that pineal tissues contained no centrosymmetric crystals, thus proving the presence of piezoelectricity. Both mulberry-like and faceted crystalline calcifications were observed by scanning electron microscopy (SEM). The calcite micro- crystals in the pineal gland when stimulated through exercises of deep breath (as per guided meditation) they start vibrating, creating friction and give off light; the phenomenon is known as piezo luminescence.

We can understand when attaining the so-called illumination/opening of the third eye by a couple of signs that differ from person to person. Some people might experience sensations at the spot, such as vibrations or pressure and you might also start having more vivid dreams or feel more connected to your intuition.

Building Strong and Supportive Relationships and Friendships

I emphasize the importance of building nurturing, healthy relationships that align with our values and aspirations since through these connections we find the strength, inspiration, and motivation to overcome obstacles and achieve our goals. By investing in our relationships and seeking self-improvement through self-awareness, we unlock our true potential as powerful and complete human beings. We are social being and we need to interact with one another but it is important to prioritize with whom we shall spend more time and most time of our lives. As seen in our day-to-day societies, the majority of us spend our time at work, or at school with strangers that we call colleagues. And the remaining time after work or study we have to spend with family and friends. Well, obviously many times this is it is not enough and those relations start to crack. And we end up being caught in this tumultuous turmoil with a mixture of thoughts and emotions that can lead to stress and further develop to many other facets! We need to STOP and BREATHE and then take the journey within.

When we realize what it is that we want and seek, it is essential to invest our time in building healthy, strong and supportive relations – friendships. *"Tell me who your friends are and I'll tell who you are".* It is believed to be an old Spanish saying that means that you can predict a person's behavior by analyzing the people they hang out with.

The relations we have with the Self, with the Creator, with our parents, our children, our friends define the quality of our life. What gets in the way of relations is our state. And problems and pain appear when there is conflict:

Conflict comes out of meaning. Meaning = Emotion. Emotion = Life. The meaning we give the other's words, behaviors and actions will make us feel and react. When we are in a good state, we automatically have good communication with all! But we often overthink or give wrong meanings and all those lead us to wrong conclusions. So, we must become masters of meanings in order to become masters of our lives!

All relationships play a crucial role in our personal growth, self-improvement, and overall wellbeing. And we must understand the significance of cultivating meaningful connections following our inner guidance on how to develop and maintain these valuable friendships.

First and foremost, it is important to recognize that strong friendships are not only about having someone to share fun and laughter with; they also serve as a pillar of support during life's challenges. Surrounding ourselves with like-minded people who share common goals and aspirations can provide us with the encouragement and motivation we need to navigate life's ups and downs of course, with confidence and grace.

To build and nurture such friendships, it is absolutely crucial to start with self-awareness. Understanding oneself and one's values is the foundation for attracting the right people into our lives. When we know who we are and what we stand for, we can seek out friends who align with our beliefs and aspirations. This self-awareness also enables us to be authentic in our relationships, fostering deeper connections based on trust and vulnerability.

Once we have identified the potential friends, it is important to invest quality time and effort into building these relationships. This involves active listening, empathy, and genuine interest in the lives of our friends. By being present and supportive, we create a safe space for open and honest communication, allowing for a deeper understanding of each other's joys and struggles.

In addition, reciprocity is key in maintaining strong friendships. It is essential to show up for our friends in times of need, providing a shoulder to lean on and lending a helping hand whenever possible. Likewise, we should not hesitate to express our own vulnerabilities and seek support when necessary. True friendships thrive on mutual support and understanding.

Lastly, it is important to remember that friendships, like any relationship, require effort and nurturing. Regularly checking in with our friends, spending quality time together, and celebrating their achievements are all ways of strengthening these bonds. By prioritizing our friendships and investing in their growth, we create a network of powerful and complete human beings who uplift and inspire each other, ultimately leading to our own empowerment and personal fulfillment.

But there is also another aspect I would like to bring forth, about attachments. We should never create and attach to anyone or anything, for our own sake. People are coming and going, friends stay as long as they are supposed to stay in our lives or new people may come exactly in those moments, we most need them. And, after that moment it's gone, those people are gone with it too but we can always keep them dearly in our memories. And here I should mention what is said about intimate relationships. When in a couple, people get together and they feel like they have their soul-mate or even a twin flame as a partner. And yet, this relationship may end too because, remember, the only constant thing in and about the universe is change. Nothing stays forever the same. We are all evolving, and our relationships too! So, train yourself to remain neutral, without judgment or attachments. And whatever life gives you, be grateful, for the moments of joy, for the moments of sorrow, see them as opportunities to grow. Let yourself to be happy and grateful for small things like the green grass, the smell of a flower and so on and life will smile back to you.

In conclusion, building strong and supportive friendships is an essential aspect of our life journey. By cultivating meaningful connections based on self-awareness and authenticity, investing time and effort into these relationships, and fostering reciprocity and mutual support, and respect, we create a powerful network of human beings who navigate life's challenges with confidence and grace. These friendships become a constant source of inspiration, encouragement, and empowerment, helping us become the best versions of ourselves.

Make a daily routine; every morning put yourself under energetic protection, so that no foreign influences are rich to you. Imagine that you "dress" yourself in an overall shield of protection from the top of your toes to the tip of your head. Another good technic is to say, "nothing from what I don't need, gets to me!" Or you can use my personal technic: you imagine yourself as being surrounded by a sphere of light with you in the center, saying out loud *I am a clean, perfect channel of light for the Divine Source of all creation, I receive, I accept and I am grateful for the complete health and prosperity, every day of my life!*

Establishing Healthy Communication Patterns

Effective communication is a crucial aspect of personal growth and empowerment.

Communication is not just about the words we speak; it encompasses the entire process of conveying our thoughts, feelings, and intentions. Developing effective communication skills is essential for building strong relationships, expressing ourselves authentically, and navigating life's challenges with confidence and grace. Our words are meant to express our thoughts and emotions. They are not just mere combinations of letters but energy blocks that build and shape our reality. Hence the importance of what we say, not only of what we think and feel! If we are what we think, we become what we mostly speak about. If we keep a positive attitude and use proper words, affirmations, through daily repetition we can manifest a positive outcome in our lives. The story that we tell ourselves is a compendium of words mirroring our beliefs. So, once you believe something, your brain retains it and looks for those things to bring to you in order to confirm your belief in yourself, to validate and to enforce it!

Remember though that words and affirmations alone without sustained action is the beginning of delusion! More efficient and with much impact is the incantation which is when you say something out loud with the certainty of believing it, over and over and over again with passion! Through this technic you actually activate your nervous system into it!

Suggestion:

Do ten days of incantations with a vibratory voice while looking in the mirror and watch the results: **Every day, in every way I am getting healthier, happier, stronger, richer (or whatever else you may like to add)!** And start living it, day by day, every day a bit more.

One key aspect of healthy communication is self-awareness. But before we can effectively communicate with others, we must first understand ourselves. Take the time to reflect on your emotions, beliefs, and values. Recognize how they influence your communication style and be open to growth and change.

Another important element of healthy communication patterns is active listening. Truly listening to others involves giving them your full attention, without judgment or interruption. Practice empathy and try to understand their perspective, even if you disagree. By doing so, you create a safe space for open and honest dialogue. Listening to your internal voice is also very important so that you take the correct decisions in life! And this voice comes without words it is pure silence, just feeling!

Assertiveness is another crucial trait in healthy communication. It involves expressing your needs, wants, and most importantly, boundaries in a clear and respectful manner. Learning to assert yourself empowers you to advocate for your own wellbeing and ensures that your voice is heard.

In addition to assertiveness, effective communication also requires effective conflict resolution skills. Conflicts are inevitable in any relationship, but how we handle them determines the outcome. The best way for this is to learn to approach conflicts with empathy, active listening, and a willingness to find common ground. Avoid blame and defensiveness, and instead, focus on finding mutually beneficial solutions. Try to explain your needs by addressing them in relation to you not to the other like we usually do. For example, when you want to tell your partner or friend about what you don't like, as we mostly do, try to twist the perspective from negative to positive, on one hand. On the other hand, if you insist and want to say about what you dislike try to address the issue by referring to you as the one that cannot contain the behavior of the partner not by complaining about what we don't like to the other person. This way makes them react and changes the outcome.

Furthermore, establishing healthy communication patterns involves setting healthy boundaries. Boundaries protect our emotional and physical wellbeing, and they play a vital role in maintaining healthy relationships. Identify your limits and communicate them assertively and respectfully. Remember, setting boundaries is not selfish; it is an act of self-care and self-respect.

Lastly, practice self-compassion throughout your communication journey. Communication is a skill that requires practice and patience. Treat yourself with kindness and understanding as you navigate through the challenges and successes of establishing healthy communication patterns.

By consciously developing healthy communication habits, you will not only enhance your relationships but also strengthen your self-esteem and personal growth. Embrace the power of communication to transform your life and become the empowered person you aspire to be.

Setting Boundaries in Relationships

In the journey of personal growth and empowerment, setting boundaries in relationships is an essential skill that everyone must cultivate. Whether it's with a partner, family member, friend, or colleague, establishing healthy boundaries is crucial for maintaining your self-worth, emotional wellbeing, and overall happiness.

Understanding the concept of boundaries is the first step towards building healthier relationships. Boundaries are like invisible fences that protect your personal space, values, and emotions. They act as a compass, guiding you towards fulfilling connections while safeguarding your individuality. By setting clear boundaries, you communicate your needs, expectations, and limits, fostering mutual respect and understanding and keeping yourself in a healthy state of mind.

It is very important to understand the various aspects of setting boundaries in relationships, and to find the proper guidance specifically tailored to those seeking empowerment and personal fulfillment. The importance of self-awareness, as well as understanding your values, desires, and limits is crucial for effectively establishing and maintaining boundaries. There are practical techniques and strategies for establishing and communicating boundaries assertively yet compassionately, ensuring that your needs are met without compromising your self-esteem.

Suggestion:

You can use the technic of the cocoon: imagine surrounding yourself with a cocoon of light and feel the boundaries of it on the left, on the right, in front, behind, on top of your head, under your feet. Visualize the "walls" of the cocoon and analyze how thick or thin they are, try the consistency if it is flexible, stretchy or on the contrary rigid. See the boundaries and let yourself sense where do you feel comfortable, how far away from where you are, in the center of the cocoon, they are. Once you do it, that's your personal space, explore it, feel it, get used to it. You can always make it bigger or smaller, depending on how you feel comfortable.

But in this process the fear and guilt often associated with setting boundaries play quite a role. It highlights the significance of self-care and emphasizes that prioritizing your own wellbeing is not selfish but rather a necessary act of self-love. I encourage you to overcome the societal conditioning that often pressures human beings to prioritize the needs of others over their own.

Through insightful anecdotes, empowering exercises, and real-life case studies, this process of setting limits aims to equip you with the necessary tools to establish and maintain strong boundaries in your relationships. It invites you to embark on a journey of self-exploration, guiding you towards a deeper understanding of your desires, limits, and values. By setting boundaries, you create a space for growth, self-fulfillment, and the development of authentic connections that will contribute to your overall empowerment.

In conclusion, setting boundaries in relationships is a fundamental aspect of personal growth and empowerment. By embracing the power of boundaries, you can confidently navigate the complexities of relationships, assert your needs, and create a fulfilling life aligned with your true self.

Resolving Conflict and Repairing Relationships

In our journey towards empowerment and personal growth, it is essential to acknowledge that conflicts and damaged relationships are an inevitable part of life. However, the way we handle these challenges can make all the difference in our quest to become powerful and complete human beings. The following steps may help us navigate with grace and confidence our inner journey of self-discovery:

1. *Cultivating Self-Awareness*: conflict resolution begins with understanding our own emotions, needs, and triggers. By developing self-awareness, we can stay grounded and

respond rather than react impulsively. This self-improvement journey requires introspection, mindfulness, and a willingness to explore our own strengths and weaknesses.

2. *Effective Communication*: open and honest communication is the cornerstone of resolving conflicts and repairing relationships. We can use active listening techniques, assertiveness, and nonviolent communication strategies that empower us to express our needs and concerns while respecting the perspectives of others.

3. *Conflict Resolution Techniques*: conflict often arises from differing opinions, values, or expectations. Using various conflict resolution techniques, such as compromise, collaboration, finding win-win solutions and focusing on mutual respect and understanding, we can transform conflicts into opportunities for growth and connection. Only when we say something that's not arguable the message is received. And yet, when we are having a conflict and we are upset and angry we can still communicate with people without creating a further argument by using a technic proposed by dr. Julie Colwell called S.E.W. (Sensation. Emotion. Want.)

Sensation – describe it to the other; it is something kinesthesis so define also where you feel it.

Emotion – share it with the other; Emotion is related to time, for example: 1 minute when anger feels like an eternity!

Want – what is it that you want or don't want.

4. *Forgiveness and Healing*: repairing damaged relationships requires the courage to forgive and let go of resentment. I emphasize here the importance of forgiveness for our own wellbeing and heal emotional wounds while setting healthy boundaries.

5. *Empathy and Compassion*: developing empathy and compassion towards us and towards others is vital for resolving conflicts and rebuilding relationships. We should explore practices that cultivate empathy, allowing us to see situations from different perspectives and facilitate understanding and reconciliation.

6. Seeking Support: Sometimes, conflicts can be complex and overwhelming. It is important to seek support from trusted friends, mentors, or professional coaches who can provide guidance and help us navigate challenging situations with wisdom and grace.

Through self-improvement, self-awareness, and a commitment to growth, we can transform conflicts into opportunities for personal development and create stronger, more meaningful connections with others. Remember, conflict is not a sign of weakness; it is an invitation to grow, heal, and become the powerful and complete being you aspire to be.

CHAPTER 7

Balancing Life's Demands: Prioritizing Self-Care and Wellness

Recognizing the Importance of Self-Care

In the journey towards empowerment and personal growth, one often overlooks the vital role that self-care plays. On our life journey, striving to become powerful and complete, it is crucial to recognize the significance of prioritizing our own wellbeing.

Self-improvement through self-awareness and life coaching requires a solid foundation of self-care. It is impossible to navigate life's challenges with confidence and grace without taking care of our physical, mental, and emotional health. This subchapter serves as a gentle reminder to place ourselves at the top of our priority list, allowing us to show up fully in all aspects of life.

Self-care encompasses a variety of activities and practices tailored to suit individual needs. From carving out time for relaxation and rejuvenation to establishing healthy boundaries and practicing mindfulness, self-care provides the necessary fuel to thrive.

Many people tend to put others' needs above their own, often neglecting themselves in the process. Therefore, I would like to emphasize the importance of self-care as an act of self-love and self-respect. By recognizing that we deserve care and attention, we cultivate a sense of worthiness and empowerment, enabling us to face life's challenges head-on.

The problem we are confronting and is common to the whole population of Earth is that we are thinking and over thing more than feeling! So, try instead to feel how a person makes you feel? Stop thinking and let your feelings "tell" you! Let yourself return to your natural nature.

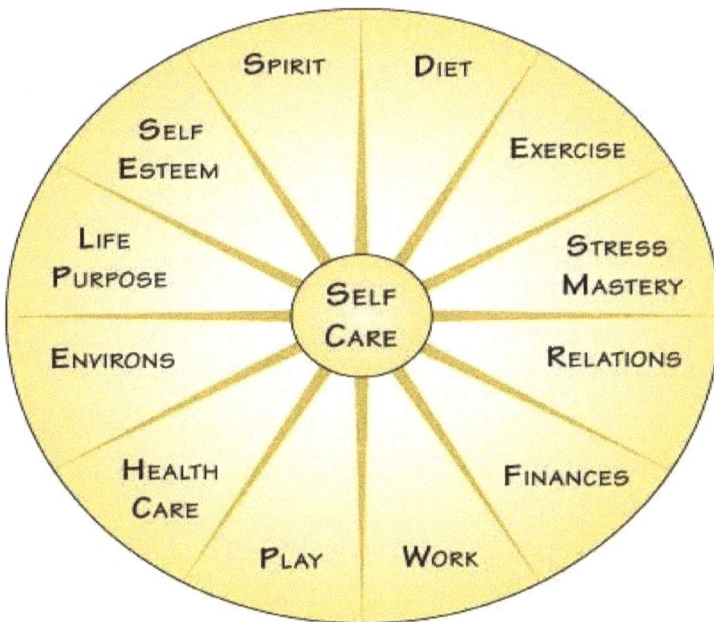

THE CIRCLE OF LIFE

Sport, engaging exercises and thought-provoking anecdotes, encourages people to reflect on their current self-care practices and make necessary adjustments. You should use practical strategies to incorporate self-care into your busy schedules; overcoming the common misconception that self-care is a luxury reserved for those with abundant free time. For example, we don't have to exercise 3-4 hours a day, 20-30 minutes are more than enough if done efficiently and with a focus to our breath. There are people who train 20's alternating intense work with slower workout while breathing through an oxygen mask. This technic proves to bring some interesting outcomes!

By recognizing the importance of self-care, people can experience a profound transformation in their lives. And self-care is not selfish but rather an essential component of our journey towards becoming healthy, powerful and complete. Armed with a renewed sense of self-worth and a commitment to self-nurture, we can navigate life's challenges with the confidence and grace we deserve.

In conclusion, embracing self-care as a fundamental aspect of our journey, we can unlock our true potential and embark on a transformative path towards becoming our most empowered selves.

Developing a Personalized Self-Care Routine

In today's fast-paced world, it's easy to get caught up in the chaos and demands of daily life. While striving to become powerful and complete individuals, we often neglect ourselves in the process. However, self-care becomes a necessity for our overall wellbeing. To navigate life's challenges with confidence and grace, we must prioritize ourselves and develop a personalized self-care routine.

Self-improvement through self-awareness and life coaching forms the foundation of our journey towards empowerment. By taking the time to understand ourselves, our desires and our limitations, we can create a personalized self-care routine that truly serves our needs.

The first step in developing a personalized self-care routine is self-reflection. So, take a moment to assess your current state of wellbeing. What areas of your life are thriving, and which ones require attention? Are you neglecting your physical health, emotional wellbeing or spiritual growth? By identifying these areas, you can begin to shape your self-care routine accordingly.

Next, explore the various self-care practices available to you. Remember, self-care is not limited to bubble baths and facemasks; it encompasses activities that nourish your mind, body, and soul. So, engage in activities that bring you joy and relaxation, such as practicing yoga, journaling, or spending time in nature. Experiment with different practices until you find what resonates with you.

It's important to note that self-care is not a one-size-fits-all approach. What works for someone else may not work for you. Therefore, it's crucial to listen to your intuition and customize your routine accordingly. Your self-care routine should be tailored to your unique needs and preferences, allowing you to recharge and rejuvenate.

Additionally, consistency is key when it comes to self-care. Make a commitment to yourself and prioritize your wellbeing. Schedule self-care activities into your daily or weekly routine and hold yourself accountable. Remember, by taking care of yourself, you are better equipped to handle life's challenges and support those around you.

In conclusion, developing a personalized self-care routine is an essential step towards becoming a strong and fulfilled person. Through self-awareness and life coaching, we can identify our needs and create a routine that nourishes our mind, body, and soul. By prioritizing self-care, we can navigate life's challenges with confidence and grace, ultimately achieving a sense of power and completeness.

Nurturing Physical Health and Well-being

In the journey of empowerment and personal growth, it is essential to prioritize and nurture our physical health and wellbeing. As powerful and complete human beings, we understand the significance of taking care of ourselves holistically, and that includes our physical bodies. This subchapter is dedicated to guiding you towards achieving optimal physical health and wellbeing, allowing you to navigate life's challenges with confidence and grace.

Self-improvement through self-awareness and life coaching begins with recognizing the importance of a healthy lifestyle. Our bodies are our temples, and it is crucial to treat them with love and respect. This means implementing healthy habits such as regular exercise, balanced nutrition, and sufficient rest. Sleep is very important but the quality of it makes the difference: many people complain about waking up in the morning more tired than in the evening when they went to sleep! So, in this cases, provided health checkups are all fine, we can help improve our sleep quality by listening specific tunes during the night, like the brain waves from alpha to theta and gamma waves, guided sleep meditations, or the solfeggio sleep frequencies (all 9 frequencies: 174Hz, 285 Hz, 396 Hz, 417 Hz, 528 Hz, 639 Hz, 741 Hz, 852 Hz, 963 Hz).

Also, engaging in physical activities boosts our energy levels but also enhances our mental wellbeing, reducing stress and anxiety. Whether it is yoga, dancing, or simply going for a walk, finding an activity that brings you joy will make it easier to incorporate regular exercise into your routine.

Nutrition plays a vital role in supporting our physical health. Therefore, is advisable for us to develop a mindful relationship with food, focusing on nourishing our body rather than restrictive diets. A balanced diet consisting of whole foods, fruits, vegetables, lean proteins, and healthy fats will provide the necessary nutrients to fuel your body and mind. Remember, self-care also means allowing yourself occasional indulgences, as long as they are enjoyed in moderation. Losing weight shouldn't become a struggle when we keep being self-aware. The body is always communicating to us what it wants if we know to pay attention to its signals and learn to listen to it! In the end we are what we eat and the expression of our thoughts and emotions! So, if we don't like the image, we see in the mirror than the best thing to do is to look for things that we can change about ourselves from inside out, like our diet, our thoughts, our emotions!

While physical health is crucial, mental and emotional wellbeing cannot be overlooked. Stress and negative emotions can take a toll on our overall health. Incorporating stress-management techniques such as meditation, deep breathing exercises, or journaling can significantly improve our mental and emotional wellbeing. Additionally, seek support from loved ones or consider working with a professional life coach who can provide guidance and help you develop effective coping strategies.

Embracing self-care rituals is an essential part of nurturing physical health. **Carve out time in your schedule for activities that bring you joy, whether it's taking a relaxing bath, enjoying a good book, or practicing mindfulness.** Remember that self-care is not indulgence but a necessity for your overall wellbeing.

By prioritizing your physical health and wellbeing, you are taking a significant step towards becoming a powerful and complete person. Embrace the journey of self-improvement and self-awareness knowing that you are worthy of investing time and effort into your physical wellbeing. As you nurture your body, on all its levels, you will find that you are better equipped to face life's challenges with confidence and grace.

Cultivating Mental and Emotional Wellness

In the journey towards empowerment and personal growth, we cannot overlook the significance of mental and emotional wellness. The ability to navigate life's challenges with confidence and grace requires a solid foundation of self-awareness, emotional intelligence, and a healthy mindset. This subchapter aims to offer valuable insights and practical tools for those who are dedicated to becoming powerful and complete individuals.

The first step towards cultivating mental and emotional wellness is self-awareness. It involves developing a deep understanding of your thoughts, emotions, and behavioral patterns. By becoming aware of your strengths, weaknesses, and triggers, you gain the power to consciously make choices that serve your best interests. Through various exercises and inner reflections, you can explore your inner landscape and uncover hidden beliefs that may be holding you back from reaching your full potential. Only through an inner journey we can discover our real selves and understand what we must do about it!

Emotional intelligence is another crucial aspect of mental and emotional wellness. It encompasses the ability to recognize, understand, and manage our own emotions, as well as the emotions of others. By enhancing our emotional intelligence, we can improve our relationships, communication skills, and overall wellbeing.

Furthermore, cultivating a healthy mindset is essential for achieving personal empowerment. It involves adopting positive thinking, resilience, and self-compassion. We can explore the power of affirmations, or gratitude practices, and reframing negative thoughts. Remember, where attention goes their energy flows! So, by learning to rewire your mind, you can overcome self-doubt, embrace change, and cultivate a growth mindset that propels you towards success.

The inner journey helps us awaken to self-awareness, emotional intelligence, and a healthy mindset, revealing also the importance of self-care and stress management. Using practical tips for incorporating self-care rituals into our daily routine, such as mindfulness exercises, relaxation techniques, and the importance of setting boundaries will transform us into powerful and complete beings. By prioritizing self-care, we actually nourish our mental and emotional wellbeing, allowing ourselves to approach life's challenges with a clear and focused mindset. Becoming aware of who you really are and staying focused leads us to live a fulfilled life!

In conclusion, cultivating mental and emotional wellness is a fundamental aspect of personal empowerment. By deepening your self-awareness, enhancing your emotional intelligence, adopting a healthy mindset, and prioritizing self-care, you can navigate life's challenges with confidence and grace. **Remember, the journey towards empowerment begins within** but it is for those seeking to empower themselves and achieve a sense of completeness!

Creating Work-Life Balance for Sustainable Success

In today's fast-paced and demanding world, achieving a harmonious work-life balance has become essential for all of us who seek to become powerful and complete individuals.

Work-life balance is not about dividing our time equally between work and personal life; rather, it is about finding a sustainable equilibrium that allows us to thrive in all areas of our lives. And again, it begins with self-awareness and understanding our priorities and values. By identifying what truly matters to us, we can make conscious choices that align with our aspirations and create a fulfilling life.

One of the first steps towards achieving work-life balance is setting boundaries. Learning to say no when necessary and establishing clear limits in both our professional and personal lives is crucial. Setting boundaries and using tips on how to communicate them assertively and ensuring that our time and energy are dedicated to what truly matters are useful tools towards attaining this balance.

One suggestion that is proven to work miracles comes from the well-known story of the two friends and work colleagues John and Bill. They were going together to work and from work back home, in Bill's car. Every morning while waiting for his friend, Bill notices that every time John spends a couple of minutes with his family to hug and kiss, wishing to each other a wonderful day and when he drives him back home, John never enters home before passing his hand through the leaves of the tree in front of his house. So, curious to know the reason, Bill asks John why he is doing all these gestures, what is the meaning of all these. The answer he gets is so inspiring and effective: in the morning John's family hugging and wishing for good day is empowering each of them for the day to come while in the evening, John never enters home full of the stressful energies collected during the day, he is releasing those to the tree by passing his hand through the leaves and, by practicing his intention of liberating himself, he transfers all those unnecessary thoughts and energies to the tree. Like this he cleanses himself and is ready to join his family, full of love and good state of mind, allowing himself to have a quality time with his beloved!

So, we can use practical strategies to manage our time effectively, prioritization techniques, the art of delegating, and the art of saying yes to the right opportunities and no to the things or people we don't need in our life. By mastering time management skills, we can optimize our productivity and create space for our personal lives, passions, and relationships.

Balancing family with work, with hobbies, friends and free time for ourselves, we can live our lives with grace and fulfillment. All the above areas are equally important in our lives and if one of these is not met then issues arise and solutions are found exactly in identifying and developing that specific area! Creating a balance between all of the above areas is a continuous process that requires ongoing reflection, self-awareness, adjustment, and self-compassion. By prioritizing work-life balance, we can cultivate sustainable success and become the powerful and complete individuals we aspire to be.

CHAPTER 8

———— ◆◆ ◆ ◆◆ ————

Embracing Change and Resilience:
Thriving in Times of Transition

Understanding the Nature of Change

Change is an inevitable and constant part of life. It can be both exciting and terrifying, as it forces us to step out of our comfort zones and embrace the unfamiliar. In this chapter we delve into the essence of change and how it can be harnessed to propel us towards personal growth and fulfillment.

Change can manifest in various ways - from major life transitions such as career shifts or relationship changes to smaller, everyday adjustments. Regardless of its scale, change is a catalyst for self-discovery and empowerment. By understanding the nature of change, we can navigate its twists and turns with confidence and grace.

One crucial aspect of change is recognizing that it is a process, not an event. It unfolds gradually and requires our active participation. This realization empowers us to take control of our own destinies, as we understand that change is not something that happens to us, but something we actively engage in. Embracing this perspective allows us to approach change with a proactive mindset, ready to make conscious choices and take intentional actions.

The only constant in the universe is change!

Another vital element in understanding the nature of change is acknowledging that it often brings discomfort and uncertainty. Change disrupts the familiar patterns we have grown accustomed to, challenging our beliefs and pushing us out of our comfort zones. However, it is in these moments of discomfort that true growth occurs. By embracing discomfort and viewing it as an opportunity for learning and expansion, we can tap into our inner strength and resilience.

Understanding the nature of change is a cornerstone of personal growth and empowerment. By embracing change as an opportunity for self-discovery and transformation, you will pave the way towards becoming a powerful and complete person who navigates life's challenges with confidence and grace.

In this subchapter, we will explore practical strategies to navigate change effectively. We will delve into the power of self-awareness, helping you cultivate a deep understanding of your values, strengths, and aspirations. Through exercises and reflection, you will gain clarity on your vision for the future and identify the steps needed to get there.

Implementing change as a tool to evolve implicates strategic approaches that promote personal growth, adaptability, and continuous improvement. Here are some strategies that can be employed to harness the power of change for evolution:

1. Adopt a Growth Mindset

- Embrace Challenges: View challenges as opportunities to learn and grow rather than obstacles. This mindset fosters resilience and adaptability.
- Learn from Criticism: Use feedback, even when critical, as a tool for improvement. This encourages continuous self-assessment and development.
- Celebrate Effort Over Results: Focus on the process and effort put into tasks rather than just the outcome, promoting a love for learning and perseverance.

2. Set Clear and Flexible Goals

- SMART Goals: Establish Specific, Measurable, Achievable, Relevant, and Time-bound goals to provide direction and motivation.
- Reassess and Adjust: Regularly review and adjust goals based on new information and changing circumstances to stay aligned with evolving aspirations.

3. Develop Resilience and Coping Mechanisms

- Stress Management Techniques: Practice stress-reducing activities like mindfulness, meditation, or exercise to build resilience against change.
- Problem-Solving Skills: Enhance your ability to tackle challenges methodically and creatively, turning problems into opportunities for growth.

4. Cultivate a Supportive Environment

- Build a Network: Surround yourself with supportive and like-minded individuals who encourage growth and provide constructive feedback.

- Seek Mentorship: Find mentors who can offer guidance, share experiences, and provide insights to help navigate changes effectively.

5. Continuous Learning and Skill Development

- Lifelong Learning: Engage in continuous education through formal courses, workshops, or self-directed learning to stay adaptable and knowledgeable.
- Skill Diversification: Develop a broad set of skills that can be applied in various contexts, enhancing flexibility and employability.

6. Embrace Adaptability and Flexibility

- Stay Open to New Experiences: Be willing to step out of your comfort zone and try new things, which can lead to unexpected growth opportunities.
- Adapt to Change: Develop the ability to quickly adapt to new situations and environments, which is crucial for personal and professional evolution. Be flexible, allowing and embracing change

7. Reflect and Self-Assess Regularly

- Journaling: Maintain a journal to reflect on experiences, track progress, and identify areas for improvement.
- Regular Self-Assessment: Periodically evaluate your strengths, weaknesses, and growth areas to stay aware of your developmental journey.

8. Implement Incremental Changes

- Small Steps: Break down larger goals into smaller, manageable steps to make the process of change less overwhelming and more achievable.

- Consistent Progress: Focus on making consistent, incremental improvements rather than seeking immediate, large-scale transformations.

9. Use of Technology and Tools
- Use Apps and Tools: Utilize productivity apps, online courses, and other digital tools to facilitate learning and manage change effectively.
- Stay Informed: Keep up with technological advancements and trends that can impact your field and open up new growth opportunities. Read every day to keep the entrainment of your brain.

10. Emphasize Health and Wellbeing
- Physical Health: Maintain a healthy lifestyle through regular exercise, proper nutrition, and adequate sleep to support overall wellbeing.
- Mental Health: Prioritize mental health by seeking professional help when needed, practicing self-care, and engaging in activities that promote emotional wellbeing. Practice meditation and mindfulness

11. Practice Patience and Persistence
- Long-Term Perspective: Understand that meaningful change and evolution take time and stay committed to your goals despite setbacks.
- Celebrate Milestones: Acknowledge and celebrate small achievements along the way to stay motivated and recognize progress.

Suggestion: practical application example

Imagine you want to evolve your career. Implementing these strategies might look like this:

- Set SMART Goals: Define clear career objectives and the steps needed to achieve them.
- Lifelong Learning: Enroll in courses to acquire new skills relevant to your desired career path.
- Build a Network: Attend events specific to the industry or area of expertise, to connect with professionals and potential mentors.
- Adaptability: Be open to new roles or projects that push you out of your comfort zone.
- Reflect and Assess: Regularly evaluate your career progress and adjust as needed.

By employing these strategies, you can effectively use change as a tool to evolve, fostering continuous growth and adaptability in both personal and professional realms.

Embracing Change as an Opportunity for Growth and Resilience

In the journey towards empowerment and self-improvement, one of the key lessons to learn is the importance of embracing change as an opportunity for growth. Change, though often feared and resisted, can be a powerful catalyst for personal transformation and fulfillment. As human beings seeking to become powerful and complete, we must understand that change is not something to be feared, but rather embraced as a steppingstone towards our fullest potential.

Change can manifest in various forms - from major life transitions to subtle shifts in our mindset and perspective. It is through these changes that we are able to shed old patterns and beliefs that no longer serve us and make room for new experiences and opportunities. By embracing change, we open ourselves up to a world of possibilities and unlock our true potential.

To navigate change with confidence and grace, it is crucial to cultivate self-awareness. Self-awareness allows us to understand our emotions, thoughts, and behaviors in relation to the changes happening in our lives. It enables us to identify areas for growth, confront our fears, and make conscious decisions about how we want to respond to change.

Professional coaching can be a powerful tool in this process, providing guidance and support as we navigate the challenges and uncertainties that change brings. A skilled coach or therapist can help us explore our fears and limiting beliefs and develop strategies to overcome them. They can also assist us in setting meaningful goals, creating action plans, and staying accountable to ourselves throughout the change process.

When we embrace change as an opportunity for growth, we cultivate resilience and adaptability. We learn to trust in our ability to navigate the unknown and emerge stronger on the other side. We become more open to new experiences and perspectives and develop a greater sense of self-confidence and empowerment.

Remember, change is not something to be feared or avoided, but rather embraced as a natural part of life's journey. By choosing to view change as an opportunity for growth, we empower ourselves to step into our full potential. So, let go of resistance, embrace the unknown, and trust in your ability to navigate change with confidence and grace.

There are many practical and efficient techniques that offer valuable tools for developing resilience. By exploring and applying various strategies such as reframing limiting beliefs, identifying and leveraging personal values, and setting realistic goals we can empower ourselves to overcome obstacles and embrace change, building resilience in the process.

Uncertainty often triggers fear and anxiety, making it challenging to stay resilient. The fear of the unknown or the unfamiliar can be very frustrating and it is mostly holding us back. By reframing our perception of uncertainty and embracing it as an opportunity for growth, we can transform fear into fuel for resilience. Fear must be lived one way or the other but in the sense of being resilient and allowing yourself to embrace the things that you hold yourself restrained, with kindness and grace.

Navigating uncertain times can be emotionally and physically draining, making self-care an essential practice. Managing stress, establishing healthy boundaries, and fostering self-compassion, enables us to recharge and replenish our resilience reserves.

Ultimately, we should understand that resilience is an ongoing journey, not a destination. It is through perseverance and self-reflection that we can embrace setbacks as opportunities for growth, celebrate progress, and cultivate an unwavering belief in the ability to overcome any challenge that comes our way.

Adapting to Life Transitions with Grace

Life is a journey consisting of a series of transitions, and how we navigate these changes can greatly impact our growth and personal development. As empowered human beings, it is essential that we approach these transitions with grace and confidence, leveraging them as opportunities for self-improvement and self-awareness. In this subchapter, we will explore effective strategies and mindset shifts to help you embrace life's transitions with poise and determination.

1. Embrace the Unknown: Life transitions often bring uncertainty, and it is crucial to embrace the unknown rather than

resisting it. By accepting that change is a natural part of life, you open yourself up to new possibilities and experiences. Embracing the unknown allows you to grow and discover your true potential.

2. Cultivate Self-Awareness: One of the most powerful tools in navigating life transitions is self-awareness. Take time to reflect on your strengths, values, and passions. Understanding yourself on a deeper level will help you make informed decisions and navigate transitions that align with your authentic self.

3. Practice Resilience: Life transitions can be challenging, but cultivating resilience will enable you to bounce back from setbacks and adapt to new circumstances. Remember that setbacks are an opportunity for growth and learning. Develop resilience by focusing on solutions, maintaining a positive mindset, and seeking support from mentors or coaches.

4. Embrace Your Emotions: Transition periods often bring a rollercoaster of emotions, and it is important to honor and embrace them. Allow yourself to feel and process your emotions without judgment. Seek healthy outlets for emotional release, such as journaling, meditation, or engaging in activities that bring you joy.

5. Create a Support System: Surround yourself with a network of like-minded human beings who are also navigating life transitions. Share your experiences, challenges, and triumphs with each other. A supportive community can provide guidance, encouragement, and different perspectives, empowering you to face transitions with confidence.

6. Set Clear Intentions: During periods of transition, it is easy to feel overwhelmed or lost. Setting clear intentions will help you stay focused and motivated. Define what you want to accomplish during this transition and break it down into

actionable steps. By setting intentions, you are actively participating in creating the life you desire.

Embrace these strategies and empower yourself to become the powerful and complete being you aspire to be.

Reinventing Yourself in the Face of Change

Change is an inevitable part of life, and it often presents us with challenges that can leave us feeling lost and uncertain about our future. However, as powerful and complete individuals, we have the ability to not only navigate through these changes but also to reinvent ourselves in the process. In this subchapter, we will explore this concept of reinvention and how it can be a transformative journey towards self-improvement and personal growth.

Reinventing yourself begins with self-awareness, which is the foundation for any successful transformation. It requires a deep understanding of who you are, what you value, and what you want to achieve in life. By taking the time to reflect on your strengths, weaknesses, and aspirations, you can gain clarity on the areas you want to improve and the goals you want to set for yourself.

Once you have identified your desires and goals, the next step is to embrace change and view it as an opportunity for growth. Change can be intimidating, but it is also the catalyst for personal transformation. By stepping out of your comfort zone and embracing new experiences, you allow yourself to discover untapped potential and unlock hidden talents.

A mentor, a guide, a life coach and even a book can provide guidance and support as you navigate through the challenges of change. These can help you identify and overcome limiting beliefs, develop a plan of action, and hold you accountable for your progress. With specific expertise, you can gain valuable insights and strategies to reinvent yourself successfully.

Reinventing yourself also requires resilience and adaptability. It is important to understand that setbacks and failures are a natural part of the journey. Instead of letting them discourage you, view them as steppingstones towards success. Learn from your mistakes, adjust your course if necessary, and keep moving forward with confidence and grace.

Lastly, surround yourself with a supportive community of like-minded human beings who are also on their journey of reinvention. Share your experiences, offer and receive support, and celebrate each other's achievements. Together, you can create a powerful network of empowerment, motivating and inspiring one another to reach new heights.

In conclusion, reinventing yourself in the face of change is a transformative journey that requires self-awareness, resilience, and a supportive community. By embracing change, setting goals, seeking guidance, and staying committed to personal growth, you can become a powerful and complete woman who navigates life's challenges with confidence and grace. Remember, the power to reinvent yourself lies within you. Take the examples of famous individuals whose life stories are so powerful and inspiring. Elon Musk is a person who's continuously eager to reinvent himself. This is evolution, this is what an empowered being aspires to, someone who is faced with harsh issues and fear finds solutions and implements them. And no matter how difficult it might be, perseveres and embraces change, with resilience and reinvents itself for a better version of the Self:

Every day in every way I'm getting better and better, the best version of myself!

CHAPTER 9

---◆◆◆◆◆◆---

Empowering Others: Making a Positive Impact on the World

Recognizing Your Influence and Impact

In the journey towards becoming a powerful and complete individual, it is essential to acknowledge and understand the influence and impact we have on ourselves, as well as on the world around us. This subchapter aims to delve into the significance of recognizing our own power, and how it can be harnessed to navigate life's challenges with confidence and grace.

Self-improvement through self-awareness is the key to unlocking our true potential. By becoming aware of our thoughts, emotions, and actions, we gain insight into the ways in which we influence our own lives. Understanding that we have the power to shape our own destiny and reality empowers us to make conscious choices that align with our values and aspirations. Recognizing our influence helps us take control of our lives and encourages personal growth.

We often tend to think about how we would like others to change in order to be more in tune with us. And our minds get unleashed into building scenarios of how or what we would like our partner, friends, and kids to be, silently wishing for them to change for our sake. But nobody will ever change if it is not by his decision! So, if you want to see any change in the other, the first person you should look to is actually yourself: change starts from within, it's an inner journey.

We are just a mirror to the person next to us, hence the ancient Essenes writings about the mirrors. These are considered tools to understand ourselves better by interpreting the reactions and interactions we have with others. Or like the old Romanian say: "tell me who your friends are so I'll tell you who you are!"

In the realm of personal and professional development, ancient wisdom often holds the keys to unlocking profound insights.

The 7 Essene Mirrors, a mystical set of principles from antiquity, offer a unique lens through which we can view and understand our interactions, reactions, and internal dialogues.

So, let's see how these mirrors can be applied in the workplace to enhance self-awareness, improve interpersonal relationships, and foster a more harmonious professional environment.

1. The Mirror of the Moment

The "Mirror of the Moment" is about how our current emotional state can influence our perception of events and interactions with others. It suggests that if we're stressed, inner tension can influence our experiences, potentially leading to more conflictual situations because we might react more negatively or be less patient than usual.

This concept implies that our emotions act as a filter through which we experience reality at any given moment. By being aware

132

of this, we can understand how much of what we perceive as external conflict may actually reflect our internal state. Recognizing this can help us manage our emotions proactively, possibly leading to more positive outcomes and interactions.

2. The Mirror of Judgment

This refers to the phenomenon where the negative traits we notice and criticize in others are often traits we possess but haven't acknowledged or dealt with within ourselves. This is a form of projection, where we actually project our own undesirable qualities onto others.

For example, being irritated by a team member's lack of commitment might indeed reflect our own issues with engagement or motivation. Perhaps it bothers us more because it's a quality we dislike in ourselves or have struggled with previously.

Understanding this mirror can lead to personal insights, helping us to address our own issues rather than focusing excessively on the faults of others. It also encourages the development of more empathy and patience, as recognizing our own flaws in others can make us more understanding and less quick to judge. This can enhance both personal and professional relationships, fostering a more supportive and cooperative environment.

3. The Mirror of What We Lost, Gave Away or Had Taken Away

This mirror refers to how we sometimes see in others the qualities or attributes that we have lost, suppressed, or perhaps never fully developed in ourselves. This concept aligns with the idea that admiration for others can often reflect our own latent or unfulfilled potential.

For example, admiring a colleague's ability to handle high-pressure situations with ease can indeed reflect a personal desire or need to cultivate more resilience and composure. This mirror suggests that the qualities we admire are not just to be envied but can be seen as indicators of what we might aspire to develop or regain in ourselves.

The process involves recognizing these qualities in ourselves, understanding why we might have lost touch with them, and actively working to reintegrate them into our own set of skills or behaviors. This can lead to personal growth and greater professional effectiveness.

4. The Mirror of Forgotten Love

The "Mirror of Forgotten Love" suggests that if we're irritated by a behavior we once admired or valued, it may be a sign that we have distanced ourselves from that trait within our own personality.

For example, if we used to appreciate the importance of taking time to socialize and build relationships at work (workplace amity) but now find such behavior by a colleague annoying, it may indicate that we have lost touch with the value we once placed on connection and social interaction in the workplace. This mirror invites us to explore why we might have deprioritized these values and to consider re-embracing them if they were once meaningful to us. It's about recognizing and potentially reviving parts of your personality that have been neglected.

5. The Mirror of Father/Mother

The "Mirror of Father/Mother" concept suggests that our early relationships, particularly with our parents, can form a template for how we interact with authority figures later in life. This includes bosses, supervisors, or any individuals in positions of power relative to us.

For example, if we have a particularly demanding boss, our response to this authority figure might be heavily influenced by our past experiences with our parents. For instance, if someone has a highly critical parent, they might be more sensitive to criticism from their boss or may interpret their boss's demands more negatively.

Recognizing this mirror involves understanding that these dynamics are playing out and reflecting on how our reactions to authority figures might be shaped by earlier life experiences. By becoming aware of this, we can work on responding to current relationships based on present circumstances rather than past experiences. This awareness can lead to healthier, more effective interactions with authority figures in the workplace.

6. The Mirror of the Dark Night of the Soul

This mirror refers to times of intense difficulty and personal challenge, which in a professional context could be job insecurity, serious conflicts within a team, or other significant workplace issues. It suggests that such difficult periods can prompt a deep reevaluation of our life and purpose. These challenges can become catalysts for change, forcing us to face our fears, reexamine our values, and emerge stronger and more resilient. It's about finding strength and insight in the face of adversity.

7. The Mirror of Self-Perception

The "Mirror of Self-Perception" reflects the concept that there can be a discrepancy between our self-image and the way others perceive us. This is about self-awareness in the context of social feedback.

If we're frequently surprised by how colleagues describe or react to us, it suggests a gap between our self-perception and the image we project. This realization can be valuable, as it invites introspection and self-examination to better understand and bridge this gap.

The process of aligning how we see ourselves with how we are perceived by others can improve our interpersonal relationships. It helps to ensure that our intentions and actions are understood as we intend them to be, which is especially important in leadership and teamwork, where clear communication and mutual understanding are key to success.

This mirror encourages personal growth by advocating for a more objective view of oneself through the lens of external feedback.

By embracing these principles, we can enhance not only our personal wellbeing but also others. Each mirror, with its unique perspective, invites us to look deeper, challenging us to grow and thrive both personally and professionally.

So, take a moment and think about it: Which mirrors do you see most often?

However, it is not just our own lives that are impacted by our actions. As social beings, we have the ability to inspire and influence those around us. Our words, deeds, and presence can make a significant difference in the lives of others. By recognizing the impact we have on others, we can use our power to uplift, support, and motivate those in our sphere of influence.

Recognizing our influence and impact is not about seeking power over others or dominating situations. It is about embracing our inner strength and understanding the positive impact we can have on our own lives and those around us. When we recognize our power, we become agents of change and catalysts for growth. Once we take the inner journey and become self-aware, we can confidently navigate life's challenges with grace, knowing that we have the ability to shape our own destiny and make a difference in the lives of others.

In conclusion, recognizing your influence and impact is a crucial step towards becoming a powerful and complete individual. By cultivating self-awareness, embracing your power, and seeking guidance from mentors and specialists, you can tap into your true potential and navigate life's challenges with confidence and grace. Remember, you have the ability to shape your own destiny and positively impact the lives of others. Embrace your influence, empower yourself, and become the person you aspire to be.

CHAPTER 10

——◆◆◆◆——

Sustaining Empowerment: Strategies for Long-Term Success

Developing a Growth Mindset

In the journey towards empowerment and personal fulfillment, one of the most powerful tools at our disposal is the development of a growth mindset. As human beings who aspire to become powerful and complete individuals, it is crucial to cultivate a mindset that embraces challenges, sees failures as opportunities for growth, and believes in the ability to learn and improve continuously.

A growth mindset is rooted in the belief that our abilities and talents can be developed through dedication, hard work, and the willingness to step out of our comfort zones. This mindset allows us to approach life with resilience, optimism, and a hunger for personal growth. It enables us to overcome obstacles, bounce back from setbacks, and constantly adapt to new circumstances.

To develop a growth mindset, it is essential to cultivate self-awareness and recognize any fixed beliefs or limiting thoughts that may be holding us back. We must challenge the notion that our abilities are fixed and embrace the idea that we have the power to change and improve. By acknowledging that we are not defined by our past or current circumstances, we open ourselves up to endless possibilities.

One way to nurture a growth mindset is through the practice of reframing. Instead of viewing failure as a sign of inadequacy, we can reframe it as an opportunity for learning and growth. By reframing challenges as stepping-stones towards personal development, we can overcome the fear of failure and embrace the lessons that come along the way.

Another crucial aspect of developing a growth mindset is adopting a lifelong learning mentality. This involves seeking out new knowledge, skills, and experiences that expand our horizons and push us beyond our comfort zones. It means being open to feedback, embracing constructive criticism, and being willing to make mistakes as part of the learning process.

Developing a growth mindset is not a one-time achievement but an ongoing journey. It requires consistent effort, self-reflection, and a commitment to personal growth. However, by embracing this mindset, we equip ourselves with the tools necessary to navigate life's challenges with confidence and grace.

In conclusion, developing a growth mindset is a powerful asset on the path to empowerment and fulfillment. By cultivating self-awareness, reframing failures, and adopting a lifelong learning mentality, we can unleash our full potential and become the powerful and complete human beings we aspire to be.

Embracing Continuous Learning and Personal Development

In today's fast-paced and ever-changing world, personal growth and development have become essential for individuals seeking to thrive and achieve their full potential.

As human beings seeking to become powerful and complete, we understand the importance of self-improvement through self-awareness. By embarking on an inner journey of self-discovery, we gain a deeper understanding of our strengths, weaknesses, passions, and values. This self-awareness becomes the foundation upon which we can build a fulfilling and purpose-driven life.

Continuous learning plays a pivotal role in personal development. It opens doors to new opportunities, expands our knowledge, and enhances our skills. By committing ourselves to lifelong learning, we cultivate a growth mindset that empowers us to adapt to changing circumstances, overcome challenges, and seize new possibilities. Whether it's through formal education, workshops, online courses, or reading, embracing continuous learning is an investment in us that yields profound returns.

Seeking mentors or guidance is another valuable tool for personal growth and development. Working with a skilled professional can provide guidance, support, and accountability as we navigate life's challenges. A coach or a mentor helps us uncover our true potential, set meaningful goals, and develop strategies to overcome obstacles. Through the coaching process, we gain clarity, confidence, and a renewed sense of purpose.

Embracing continuous learning and personal development requires commitment and perseverance. It involves stepping outside of our comfort zones, embracing new experiences, and challenging our limiting beliefs. As we grow and evolve, we become more empowered to make conscious choices that align with our values and aspirations.

By embracing continuous learning and personal development, we can unlock our true potential and live extraordinary lives. With the right tools, mindset, and support, we can navigate life's challenges with confidence and grace, becoming the empowered and complete human beings we are meant to be.

Building a Supportive Network

In the journey towards empowerment, one of the most invaluable resources is a supportive network of like-minded individuals. Surrounding yourself with people who believe in your potential, encourage your growth, and provide guidance can make all the difference in achieving your goals. It is of outmost significance to build a supportive network and to explore practical strategies to cultivate such relationships.

In our quest towards empowerment and evolution, it is essential to recognize that we are not meant to navigate life's challenges alone. We thrive when we have a tribe of individuals who understand our aspirations, challenges, and dreams. A supportive network can provide the emotional and practical support we need to overcome obstacles, gain self-awareness, and embrace personal growth. There is a rule of life regarding networking: it should be in a proportion of 33%- 33%- 33%. Meaning that we should surround ourselves with 33% of people who are above our vibratory level who we can offer and need our help, 33% of people that are at our same level of vibration and evolution and the last 33% should be of the people above our vibration, people that inspire us, like gurus, mentors, personalities of certain fields, heroes.

To begin building a supportive network, it is crucial to assess our current relationships. Identify individuals who uplift and inspire you, also those who share similar interests and goals. These can be friends, family members, colleagues, or even mentors. Seek out communities and organizations that align with your interests, such as self-improvement groups, networking events, or online forums. Surrounding yourself with people who are already on a similar journey can enhance your growth and provide valuable insights.

Once you have identified potential members of your supportive network, it is essential to cultivate these relationships with intention. Be proactive in reaching out, sharing your goals, and expressing your desire for support. Actively listen to others, offer encouragement and empathy, and be willing to reciprocate the support they provide. Remember, building a supportive network is a two-way street – it requires both giving and receiving.

In addition to cultivating relationships, it is crucial to seek out mentors or life coaches who can guide you on your journey of self-improvement. These individuals have already walked the path you aspire to, and their wisdom and experience can be invaluable. Seek out mentors who align with your values and can offer guidance in areas you wish to develop. A mentor, a guru, or a professional therapist can provide accountability, perspective, and practical strategies to help you navigate life's challenges with confidence and grace.

But do not forget about sharing with others your knowledge, helping them and guiding them to improve themselves, once they ask for your help. Become the mentor and support others, too.

In summary, building a supportive network is a vital component of personal growth and empowerment. Surrounding yourself with individuals who believe in your potential, share similar goals, and offer guidance can propel you towards a successful life. By assessing your current relationships, seeking out like-minded communities, and cultivating intentional connections, you can create a network that uplifts and supports you on your journey towards becoming a better version of your Self. Remember, you are not alone – together, we can navigate life's challenges with confidence and grace.

Practicing Gratitude and Mindfulness

In today's challenging and chaotic world, it is easy for us to become overwhelmed and lose sight of our true nature and inner strengths. However, by incorporating gratitude and mindfulness into our daily lives, we can cultivate a sense of empowerment, resilience, and inner peace.

Gratitude is a powerful tool that allows us to shift our focus from what is lacking in our lives to what we already have. By consciously acknowledging and appreciating the blessings, big or small, we invite more positivity and abundance into our lives. In the face of challenges, practicing gratitude can provide us with a fresh perspective, helping us to navigate life's obstacles with confidence and grace.

Mindfulness, on the other hand, is the practice of being fully present in the moment, without judgment or attachment. By cultivating mindfulness, we learn to observe our thoughts and emotions without getting caught up in them. This practice enables us to respond to situations rather than react impulsively, empowering us to make conscious choices that align with our values and goals. Mindfulness also allows us to develop a deeper understanding of ourselves, enhancing our self-awareness and fostering personal growth.

From keeping a gratitude journal to practicing mindful breathing and meditation, we will delve into various methods that can help us cultivate these powerful practices.

By embracing gratitude and mindfulness, we are not only creating a foundation for personal growth and self-improvement, but we are also nurturing our overall wellbeing. It is vital that we prioritize our mental and emotional health and use practices that offer a pathway to achieve balance and fulfillment.

By practicing gratitude and mindfulness, you have the opportunity to elevate your life, navigate challenges with confidence, and cultivate a sense of inner peace and joy.

Suggestion:

There are many techniques used to practice gratitude like: write down, in your journal, ten things you are grateful to have in your life now, like your health, your house, having food on the table, for your parents, and so on. I'd suggest you use this formula by Bob Proctor:

I am so grateful and happy now that .... (I am completely healthy)

I am so grateful and happy now that................. (I am having the house of my dreams) etc.

Practice this every day before going to bed or first thing in the morning for 21 days and see how things unfold for you.

Reflecting on Your Empowerment Journey

As you continue on your journey towards empowerment, it is important to take a step back and reflect on your progress to gain a deeper understanding of your growth and accomplishments. By pausing to acknowledge your achievements and lessons learned, you will enhance your self-awareness, confidence, and grace.

Self-improvement through self-awareness lies at the core of empowerment. It is through self-reflection that you can identify your strengths, weaknesses, and areas for growth. Take the time to journal or meditate on the moments when you felt most empowered. Recall the situations where you handled challenges with confidence and grace and consider what qualities and skills you utilized. Celebrate these achievements, as they are proof of your progress and capability.

However, self-reflection or self-introspection is not solely about recognizing triumphs; it also involves acknowledging the lessons learned from setbacks and failures. Embrace these moments as opportunities for growth. Consider the challenges you faced and explore how they shaped you. What did you learn about yourself? How did you navigate through adversity? Reflecting on these experiences will allow you to see your own resilience and build upon it.

Seek guidance to better understand this process from a trusted coach or mentor who can provide an objective perspective. Share your reflections with them and discuss strategies for further growth. They can help you identify any blind spots or self-limiting beliefs that may be holding you back. Through their support, you can develop a clearer path towards becoming the powerful and complete woman you aspire to be.

Remember, empowerment is not a destination but a continuous journey. Reflecting on your growth allows you to stay present and connected with yourself. It enables you to appreciate how far you have come and inspires you to keep moving forward. By nurturing self-awareness, celebrating achievements, and learning from challenges, you will navigate life's obstacles with confidence and grace.

In conclusion, take the time to reflect on your empowerment journey. Embrace self-improvement through self-awareness and seek the proper guidance. Celebrate your achievements and learn from setbacks. By doing so, you will continue to grow, becoming the better version of your Self, navigating life's challenges with confidence and grace.

Conclusion: Embracing Your Power: Thriving as an Empowered Individual

Congratulations! You have now reached the conclusion of this book. Throughout this journey, we have explored various aspects of self-improvement, self-awareness, and the importance of mentoring and life coaching, all with the aim of helping you become a better You. As we wrap up, it is essential to reflect on the key takeaways and lessons learned.

First and foremost, empowerment begins with the inner journey within, understanding and embracing your unique power. You possess incredible strengths, talents, and qualities that make you who you are. By acknowledging and embracing these attributes, you can tap into your full potential and accomplish remarkable things. Remember, your power lies within you; it is up to you to unleash it.

Self-awareness has been a recurring theme in our discussions, and for good reason. Understanding yourself, your values, and your goals is a crucial step towards personal growth. Take the time to reflect on your strengths, weaknesses, and areas for improvement. By doing so, you can make conscious choices that align with your authentic self, allowing you to thrive in all areas of life.

Seeking professional guidance is a powerful tool that helps you evolve. It has provided you with valuable tools and strategies to navigate life's challenges. From setting clear goals to developing effective communication skills, these techniques have empowered you to overcome obstacles and create a life you love. Remember that life is a continuous journey, and coaching can be a lifelong resource to guide you towards your dreams and aspirations.

As you embark on your self-discovery journey, it is essential to surround yourself with a supportive community. Connect with like-minded human beings who are also seeking personal growth and empowerment. Share your experiences, challenges, and victories. Together, you can lift each other up and create a network of powerful and complete individuals.

Lastly, remember that empowerment is not a destination but a lifelong process. Embracing your power is a continuous endeavor that requires dedication and self-compassion. Be patient with yourself, celebrate your progress, and never underestimate the incredible potential within you.

In conclusion, as we came to the end of our journey, I hope that in this book you have found valuable inspiration, insights, and guidance to help you on your path towards greater self-awareness and authenticity to thrive as a powerful human being. Embrace your power, practice self-awareness, stay present in the moment and use different coaching techniques, to create a life filled with confidence, joy, and fulfillment. You deserve it! Remember, you have the power to accomplish anything you set your mind to. Claim it, own it, and let your light shine brightly in the world. You are a powerful and complete individual, and your journey towards empowerment has only just begun.

And, as you close the pages of this book, take a moment to reflect on your own path. The insights and strategies shared within these chapters are not just theoretical concepts; they are tools designed for you to embrace your unique journey. Life's challenges can feel overwhelming but remember that each obstacle is an opportunity for growth and transformation and a life experience.

Allow the information you've learned here to resonate in your daily life. Embrace the courage to face uncertainty and the confidence to trust in your own resilience. As you move forward, let this book serve as a companion and a reminder that you possess the strength to navigate whatever comes your way. Your inner journey is ongoing, and with each step, you are crafting a narrative of empowerment and hope. Carry these reflections with you and know that you are not alone on this journey. The power to change and thrive resides within you.

In the end, I would like to express my deepest gratitude to you, the reader, for joining me on this transformative journey. Your dedication to self-improvement and personal growth is truly inspiring, and I am honored to have been a part of your quest for self-awareness. Remember that the power to change and grow lies within you, and I have no doubt that you will continue to evolve and thrive on your path to self-discovery.

Thank you for allowing me to be a part of your journey. May you continue to explore, learn, and grow in all aspects of your life!

Wishing you happiness, fulfillment, and harmony as you continue on your path to self-discovery.

With heartfelt thanks and warmest regards,
Loredana Climena
Wonderful life journey beautiful Souls! Till we meet again!

www.ingramcontent.com/pod-product-compliance
Lightning Source LLC
Chambersburg PA
CBHW052012030426
42334CB00029BA/3185